Always Amongst Friends

The Cardiff and County Club
1866-2016

Always Amongst Friends

The Cardiff and County Club
1866-2016

Andrew Hignell

Welsh Academic Press
Cardiff

Published in Wales by Welsh Academic Press, an imprint of

Ashley Drake Publishing Ltd
PO Box 733
Cardiff
CF14 7ZY

www.welsh-academic-press.wales

First Edition – 2017

ISBN
Limited Edition hardback: 978-1-86057-127-5
Paperback Edition: 978-1-86057-129-9

© Ashley Drake Publishing Ltd 2017
Text © Andrew Hignell 2017

The right of Andrew Hignell to be identified as the author of this work has been asserted in accordance with the Copyright Design and Patents Act of 1988.

All rights reserved. No part of this publication may be reproduced, stored in a retrieval system, or transmitted, in any form or by any means without the prior permission of the publishers.

British Library Cataloguing-in-Publication Data.
A CIP catalogue for this book is available from the British Library.

Every effort has been made to contact copyright holders. However, the publishers will be glad to rectify in future editions any inadvertent omissions brought to their attention.

Ashley Drake Publishing Ltd hereby exclude all liability to the extent permitted by law for any errors or omissions in this book and for any loss, damage or expense (whether direct or indirect) suffered by a third party relying on any information contained in this book.

Typeset by Replika Press Pvt Ltd, India
Printed by Akcent Media, Czech Republic

Contents

Patron's Foreword viii
President's Welcome ix
Acknowledgements x
Preface xi

1. Brunel and the Marquess of Bute — 1
2. Mr. Heard and the Hotel Club — 11
3. The Hunting Set — 25
4. On the Move — 32
5. Good Old Frank — 40
6. A Permanent Home in Westgate Street — 48
7. Berkeley and the Brains — 59
8. Chukkas and Putters — 69
9. 1905 and all that — 77
10. The Great War — 86
11. Footballer, Father and Friend — 96
12. The Roaring Twenties — 103
13. The King and Freddie — 113
14. Motor Cars and Bodyline — 120
15. The Alexanders — 134
16. Austerity and Air Raids — 143
17. The Club's Garden, and What to Do With It? — 155
18. Sir Tasker and 'The Skipper' — 164
19. The Beeb — 171
20. Celebrating the Club's Centenary — 178

21.	New Neighbours on Westgate Street	185
22.	A Load of Old Rot!	192
23.	Hurrah to the Staff	199
24.	125 'Not Out'	211
25.	The Devolution Debate	221
26.	The Millennium Stadium	227
27.	Days of the Dinosaur	235
28.	Lift off! Into the Twenty-First Century	243
29.	An Agenda for Change	250
30.	Bleddyn and Dr Jack	260
31.	It's a Man's World No Longer	270
32.	Onwards and Upwards	278
Index		288

To Henry Heard and the founder members, without whom Cardiff and Glamorgan would not have such an outstanding club.

Patron's Foreword

'I very much enjoyed my visit to the Club in 2016. I was particularly impressed by the very evident sense of cheerful friendship between the members. That, of course, is the factor that distinguishes all successful clubs. Home and work probably occupy most of the time of most of the members, so that a congenial club provides the opportunity to meet and make friends with members from all sorts of other occupations and backgrounds.

The journey to a 150th anniversary is inevitably going to be a rich story full of interesting characters, dramatic incidents and amusing anecdotes, but it will also reflect many changes within local society and the business community. Quite evidently, as *Always Amongst Friends* shows, the 150 years of the Cardiff and County Club have been no different and I am delighted to have this opportunity to offer my congratulations and best wishes to all present members, and to wish the Club every success in the years to come.'

The Patron signs the new Visitors Book watched by the President.

H.R.H. The Duke of Edinburgh
Patron, The Cardiff and County Club
October 2016

President's Welcome

It was my great privilege to be President of the Cardiff and County Club during its 150th anniversary, and I would like to thank Dr. Andrew Hignell for his painstaking efforts in the compilation of this history which has required many hours of research, both in talking to members and delving into our archives.

Always Amongst Friends is a worthy successor to Arthur Weston Evans' volume published for the 125th anniversary and sets a benchmark for future publications to mark the Club's milestones in years to come.

The Club has been ably and honourably served over these years by those members who have served as its officers and by those who have dedicated their time to its numerous committees. On behalf of the Club's present and past members I would like to thank them all for the many hours they have spent on Club business. We are also most fortunate and thankful to have had such a loyal, committed and professional team of staff without whom the Club would not function.

I am certain that with the continued support of the members, our officers and staff will ensure that the Cardiff and County Club continues to flourish and that members will appreciate and maintain the high standards for another 150 years.

Club President, David Mansel Jones.

David Mansel Jones
President, The Cardiff and County Club
October 2016

Acknowledgements

A number of people have been of great assistance in the compilation of this book. My thanks, in no particular order, to the following for giving their time so freely for informal interviews or providing photographic material for this handsome publication – Anthony Alexander; David Bevan; Jeffrey Bird; Richard Bosworth; Chris Brain; Jeff Childs; Guy Clarke; John Cosslett; Bob Edwards; Alex Embiricos; Mair Forsdyke; Roger Gabb; Bill Gaskell; John Jenkins; Clive Johnson; Nick Lawrance; Professor John Lazarus; Brian Lee; Brian and Emma Lile; Adrian and Norman Lloyd-Edwards; David Mansel Jones; Tim Mathias; the late Dennis Morgan; Louise Mumford; Robbie Norris; Fiona Peel; Duncan Pierce; Ceri Preece; Gwyn Prescott; Gillian Redwood; Dennis Sellwood; Owen Sennitt; Gill Thomas; Hugh Thomas; Simon, Sara and Georgina Turnbull; Paul Twamley; David Watson James; David Webber; Andrew Williams; Professor Gareth Williams; Lawrie Williams; Jon and Penny Wooller; and Sally Young. My very grateful thanks as well to Katrina Coopey and Tony Davidson of the Local Studies Department of Cardiff Central Library; Susan Edwards and her staff at Glamorgan Archives; The Llandaff Society, The Insole Court Trust and the Llanishen Local History Society for answering my plethora of research queries.

Finally, and by no means least, my thanks to the committee of the Cardiff and County Club for inviting me to compile this book about such a magnificent organisation, to Ashley Drake of Welsh Academic Press for agreeing to publish the book, and to the team of proofreaders, including my wife Debra, for converting my drafts into a polished final manuscript.

Preface

The title of this book *Always Amongst Friends* is derived from one of the many comments I heard when talking with Club members in an attempt to encapsulate the character and function of Wales' premier private members' club. Together with a series of other quotes which adorn the start of each chapter, it is a most fitting title for this history of the Cardiff and County Club which proudly celebrated its 150th anniversary during 2016.

Throughout this century and a half, the Club has enjoyed a very close and intimate link with the sporting and business community of south Wales, with the Club, since the middle of the 20th century, literally being in the shadow of the Arms Park, and its modern manifestation The Millennium (or Principality) Stadium. Indeed, there exists today a delicate juxtaposition of the Club, full of Corinthian spirit in a Victorian building, with the modern 74,000-seater covered stadium, its artificial playing surface and every conceivable technical gadget. Like many successful marriages, this effective working relationship between the Club and the stadium is testimony to the fact that opposites attract.

Whilst the Stadium is a visible barometer of Welsh identity in the post-industrial world, to some the Club appears to be a relic from the past, and a reminder of the days when the great and good of south Wales society would wallow in gracious and gentlemanly club life. But, to paraphrase the song, *The Times They Are a-Changing* as, behind the façade of the red-brick building in Westgate Street, there have been some important (some might say radical) changes during the past few years. In November 2014, a vote was taken to admit women members, whilst Wi-Fi and other forms of modern technology now fill the offices and several function suites which are 'chock-a-block' on the days when international rugby takes place at the stadium, as well as on other occasions when the Club plays host to grand functions or special visitors.

The Club, like the stadium itself, has a common ancestry as both owe their very existence to the economic changes which swept across south Wales from the 19th century onwards. In many ways, the history of the Cardiff and County Club is also a history of the industrial and commercial change experienced by the Taff-side town. The modern metropolis of Cardiff

2016 is vastly different in nature, scope and size to the tiny market town which 200 years before was quite literally in the shadow of the Castle.

This dramatic transformation began in 1839 following the decision by the Second Marquess of Bute to open the docks at the mouth of the river. It was a decision which brought men and their money into south Wales, and the history of the Cardiff and County Club is, in many ways, a tale of their endeavours in so many diverse ways of life. The fact that the Club is now celebrating its 150th year is testament to the success of their deeds, with the Club being older than many other bastions of Cardiffian life, including the Welsh Rugby Union, the *Western Mail* newspaper and Brains Brewery to name but three organisations which still feature prominently in modern Welsh life yet are younger than the Club which still resides in the clubhouse in Westgate Street.

I hope that you enjoy this journey through a rich and heady mix of the economic, social, sporting and cultural history of Cardiff and its immediate environs, besides reading about the people who have made the Club what it is today, and what, in the words of Donald Box, a former Club Chairman, still remains "a most agreeable place where members can come to relax with friends, and not talk about business!".

Andrew Hignell
St. Fagan's, Cardiff
December 2016

1

Brunel and the Marquess of Bute

"The Club is intimately bound up with the history and wider development of south Wales since the mid-nineteenth century."

People living in Victorian Britain had much for which to thank Isambard Kingdom Brunel, including the design of great steamships, great viaducts, deep tunnels and other outstanding feats of civil engineering. Residents of Cardiff – both past and present – can also raise a glass in the memory of the great engineer because, as a young man, he was a member of a team of engineers employed by the South Wales Railway which was seeking to establish a suitable site for the town's main railway station, free from the danger posed of flooding from the River Taff, which as the maps of the period show, meandered close to the western edge of the small town.

Like the townsfolk, Brunel and the other forward-thinking engineers were aware of the hazards the Taff posed so, with the approval of the Second Marquess of Bute – the aristocratic owner of much of the land in the area, and of the embryonic docks a couple of miles or so downriver at its mouth – a plan was devised to create a straighter channel for the Taff. Between 1848 and 1853 Brunel duly oversaw the diversion of the river away from the western edge of the town with a new channel being created just to the north of Cardiff Bridge and to the west of the grounds of Cardiff Castle where the Marquess and his family resided for part of the year.

Not only did the diversion create an area of dry land at the southern end of St. Mary Street where a railway station could be built, it significantly changed the geography of the western edge of the small town, which at the time of the first population Census in 1801, had 1,870 residents almost all living, quite literally, in the shadow of the Castle. One of the most significant changes was the extension to an area of meadowland at the rear of the Cardiff Arms, the coaching inn and hotel, adjacent to the West Gate into the town. Originally, a townhouse, the Cardiff Arms had been converted into an inn during 1787 and, after its sale to the Marquess in 1803, it became 'the place' to stay whilst visiting the town. Indeed when King Edward VII, Prince of Wales, visited the town as a young boy, he stayed at the Cardiff Arms.

The small area of land at the rear of the inn was also used by the townsfolk for civic events, such as the celebrations in 1837 for the accession of Queen Victoria. But, until the work by Brunel and others, the land was poorly drained and prone to flooding. The diversion of the Taff ended these difficulties, besides enlarging the Park to some 18 acres. It duly became a popular place for recreation, and in May 1848 the *Cardiff and Merthyr Guardian* reported how the members of the town's cricket club 'met in the field a little westward of the Cardiff Arms Hotel.' Cricket was the first organised team sport in both Cardiff and Wales, with the cricket club – formed in 1819 – having previously played on land owned by Lord Tredegar at Splott Farm as well on the land belonging to the military authorities at Longcross Barracks, adjacent to Newport Road.

This engineering work by Brunel was therefore massively important for the subsequent evolution of Cardiff during the 19th and 20th centuries, as well as the creation of the Cardiff and County Club. Amongst other long-lasting benefits, it created an area of public open space adjacent to the town and led to leisure activities taking place immediately to the west of the central business district, and on land owned by the Bute Estate. The lower part of St. Mary Street also became a drier, safer and

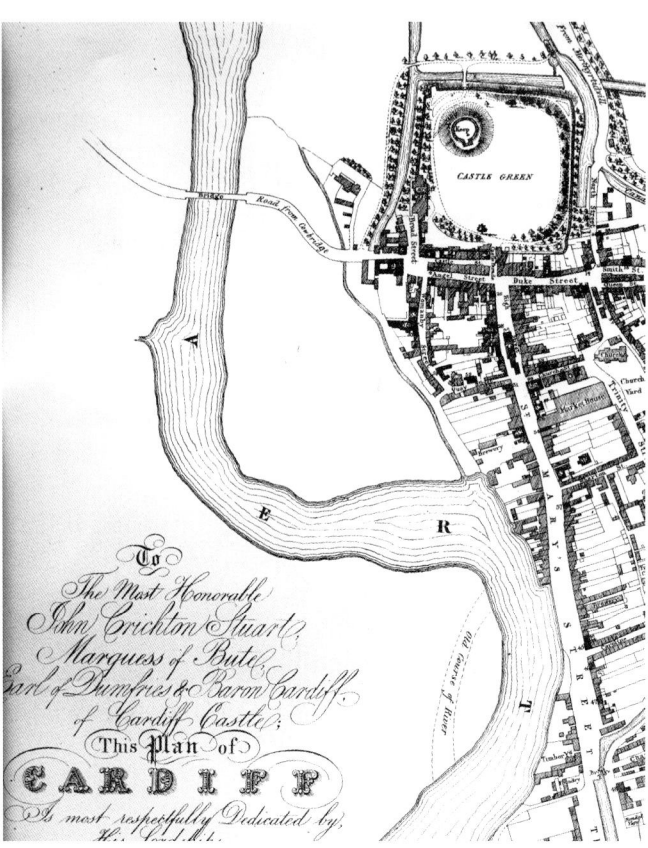

A map of Cardiff in 1830 showing the Quay and St. Mary Street (known in the 19th century as St. Mary's Street, after the 11th century church which existed near the site of the Royal Hotel until 1607, when it was destroyed by flooding).

A print of Cardiff in 1840 looking east from Cardiff Bridge. The Castle tower and that of St. John's Church dominate the skyline.

healthier place in which to live and work. Shortly afterwards a new road was created adjacent to Cardiff Arms Park – initially known as New Street before later becoming Westgate Street. Within a few years, a series of grand buildings had been erected along the eastern flank of the Street, including several more hotels, from which its guests could look across the western outskirts of the town and, if they so desired, watch the sporting activities taking place on Cardiff Arms Park.

On June 18th, 1850, the South Wales Railway opened its line between Chepstow and Swansea, followed the next year by an eastern extension to Gloucester. By this time, the number of residents in Cardiff had started to mushroom, with 18,351 being recorded in the 1851 Census. Trade was booming at the Bute Docks which had been opened in 1839, largely as a means of easily exporting the coal and iron ore from the valleys to the north, and importing the timber and other products so badly needed in the mines and industrial furnaces. Historians have rightly dubbed the Second Marquess of Bute the 'Maker of Modern Cardiff' as under his *aegis* and the shrewd actions by the administrators of his Estate, his son the Third Marquess and the Bute Docks Company, the Taffside town became a coal metropolis,

The Third Marquess of Bute.

A view of High Street looking south from the Castle in 1841, from an engraving by JH Lekeux. The river meadow alongside the Arms Park, to the right of the print, is clearly visible.

and one which in 1913 exported more coal than any other port in the world.

These actions brought fame and fortune to the Second Marquess and his descendants, besides radically transforming everything associated with the town. No longer was it a sleepy market town as, by the end of the 19th century, it was a major industrial settlement, with further docks and a steelworks at East Moors, opened in 1891. That year the Census recorded 128,915 people living in Cardiff.

Unlike many manufacturing centres which the Victorian era spawned in Britain, Cardiff also had two distinctive and geographically separate business areas – one was based at the docks, and the other centred around the old town. The docks area included the Pier Head Building – the magnificent, red-brick headquarters of the Bute Docks Company – as well as from the mid-1880s the Coal Exchange where, in 1901, the world's first-ever million-pound deal is reputed to have been agreed. Previously, the traders and agents in the coal trade had chalked up prices for their products on slates hung outside their offices or struck deals in the local pubs but, from its completion in 1886 the Coal Exchange became the base for the coal and shipping magnates, as well as others involved in the trade of industrial products to and from south Wales.

The second business hub centred around the old town and included the shops, offices and other commercial properties in St. Mary Street, Duke Street, High Street, Queen Street and the newly-built Westgate Street. It was at the junction of High Street and St. Mary Street that the Town Hall had been established 1854, with the creation of the building being an example of Cardiff's growing civic pride and identity. Another example came the following year as the town's cricket club, with the full support of the Bute Estate, arranged a two-day cricket match on the Arms Park between a side representing Cardiff and District against a team of English professionals, organised by William Clarke, the Nottingham-based entrepreneur, who staged special exhibition matches across the country. Indeed, by the mid-1850s most of the principal towns and cities in England had staged matches involving an All-England Eleven, or a United England team. All had attracted decent crowds, and gambling tents were erected where spectators could bet on how many runs each batsman might make or who would take the most catches or wickets.

The match in 1855 also saw the mercantile elite, as well as other members of the urban bourgeoisie, stepping to the crease. The Cardiff and District side, numbering 22, included a number of prominent townsmen, such as the Rev. Alfred Ollivant, the Bishop of Llandaff, the Rev. Cyril Stacey who was curate of St. John's – the only medieval church in the central area – and Captain Maher, who was in charge of the lightship in Cardiff Bay. Given the status of this game, the Cardiff officials also hired a number of guest players, including E. M. Grace – the elder brother of the famed W.G. – from the West Gloucestershire club, plus a couple of journeymen professionals.

The match would have involved considerable expense for the Cardiff club, what with meeting the guarantee which Clarke had demanded when securing the fixture, as well as paying appearance money for his England players. With entrance money also being charged, it is likely that the Bute Estate also contributed towards some of the expenses, as well as overseeing the preparation of the wicket in the Arms Park, and the creation of several enclosures, including some with overhead cover, where the great and good of Cardiff society could sit and watch the match in comfort.

This contest highlighted the rising status, prosperity and confidence of the town and its inhabitants. During the period leading up to the match, the *Cardiff and Merthyr Guardian* took great delight in reporting how 'our district players have within the last few weeks been assiduously engaged in testing sinew and muscle, by throwing the ball, plying the bat and practising the various rapid manoeuvres requisite for this vigorous and healthful game.' As it turned out, perhaps more practice was needed as the local team were dismissed for 64, with William Selby, one of the professionals, being the only batsman to get into double figures. Edmund Hinkley, another one of the hired hands, then earned his match fee by taking seven wickets as the English side secured a first innings lead of 72. Batting for a second time, the local men found runs slightly easier to acquire with Frank Stacey, the brother of Rev. Stacey, displaying his talents with the bat. Nevertheless, the England team needed only 98 to win, and the crack professionals duly cruised to their target with five wickets in hand.

Despite the defeat, the *Cardiff and Merthyr Guardian* was most complimentary about the townsmen's efforts against the cream of cricketing talent from England. It proudly proclaimed how 'the event will prove an epoch from which we shall hereafter have to date the progress of south Wales to a higher elevation in the noble art of cricketing.' Flushed by the success of the game against Clarke's team, Cardiff CC – together with the Bute Estate – arranged another grand exhibition contest in 1857, this time against a team raised by John Wisden, the Sussex professional who had created a side called the United All-England Eleven.

The so-called 'Cardiff and District XXII' for the three-day contest in August 1857 once again included several members of the local gentry, including Captain George Homfray, a leading figure with the rapidly expanding cricket club at Newport and whose family, based in Tredegar, were a scion of the great dynasty of ironmasters and landowners. Indeed, the Homfray family were closely associated with many of the flourishing industrial centres in the south Wales valleys, and were to later play a leading role with the Cardiff and County Club.

Jeremiah Homfray, the uncle of George, had moved from his native Staffordshire to open the Penydarren Ironworks in Merthyr in 1805, followed by others in Ebbw Vale. He later had a few mishaps and became bankrupt before fleeing to France to escape his creditors. In contrast, George's family met with more success with his father Samuel establishing the flourishing iron works in Tredegar, and it was here, or more precisely, at Bedwellty House, that George was born in 1830. When George was 13, the family moved to Glan Usk near Newport, with Bedwellty House becoming the home of the manager of the Tredegar Iron Works.

It was through cricket that Homfray became a good friend of George Worthington, a Worcestershire-born entrepreneur who owned the Llancaiach and Gelligaer Collieries before becoming a Director of the Penarth Dock and Railway Company. Whereas George Homfray lived with his wife at Clytha Square in the fashionable St. Woolos district of Newport, and became a leading figure with the Usk-side town's cricket club, George Worthington lived at Llancaiach House, and had lodgings in Working Street in Cardiff. As well as being a successful batsman with the Cardiff club, Worthington was a keen huntsman, riding with the Roath Court and Glamorganshire Hunts, besides being a familiar face at the annual Cardiff steeplechases where he mixed with his many friends from the sporting and business world.

Despite his fine record for the Cardiff club, Worthington met with mixed success in the two games against the wandering England XIs. In 1857 he was one of the organisers of the game and, as a result, invited Homfray to play. It was mainly for social reasons as, unlike his friend, Homfray was not an outstanding cricketer. But what he lacked in talent, he more than made up for in enthusiasm, and he tirelessly promoted local sport. Following the success of matches against the English XIs, he founded in 1859 the South Wales Cricket Club – a gentleman's XI which played annual matches against teams in the Home Counties and the West Country, besides undertaking an annual tour to London, with George Worthington being a regular face in the south Wales line-up.

Homfray did not score a run in either innings in the match at the Arms Park in 1857 against the United England team, but he was not alone as half of the 44 innings of the Cardiff players ended in ducks. Once again, Frank Stacey was among the only local batsmen to worry the English professionals, who ran out comfortable winners by the sizeable margin of 135 runs.

The 1850s were, therefore, pivotal years for the sporting gentlemen of Cardiff, with high-profile cricket matches in 1855 and 1857 against crack English opposition, besides the opening on May 30th, 1855, of Ely Racecourse where – to the delight of Worthington and the other keen huntsmen – steeplechases took place on the dairy farm owned by George Thomas to the west of the town. Thousands of spectators flocked to the course before heading back into the central areas to either celebrate their good fortune or to drown their sorrows.

These sporting events during the 1850s helped to unify the residents of the expanding town of Cardiff and, as the example of the two Georges showed, they brought together a number of well-to-do gentlemen from the business and social world of south Wales, as well as so many diverse personalities in the life of Cardiff. Perhaps it was during some of the jolly banter and bonhomie exuded in no small amount at these cricket matches or at the annual Cardiff Races in Ely that the seeds were sown for Cardiff, like so many other settlements elsewhere in Britain, to have a club where the gentlemen of the town and its environs could socialise and relax in the dignified and civilised company of their equals.

Where did gentlemen meet before the Club was established?

It is quite interesting to speculate where the gentlemen of the town of Cardiff, and those from the surrounding parishes, met to socialise before the creation of the Cardiff and County Club. Sadly, no records exist but it is well-known that the pre-Victorian town had a number of taverns and hotels where those of comfortable means could mingle.

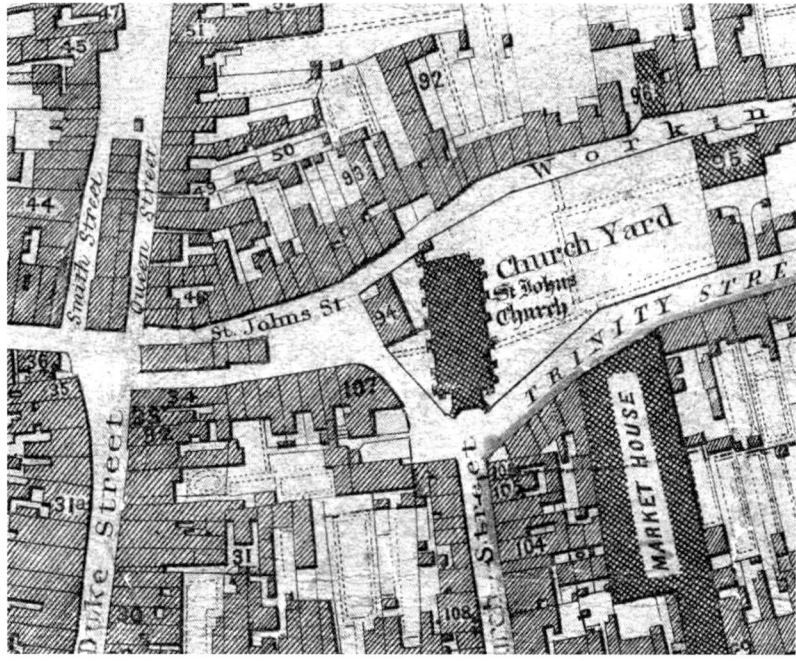

Nell's brewery was located at site 167 (opposite St. John's Church), as shown on this map from 1849 compiled by O'Rourke.

These included the Rummer Tavern in Duke Street, whilst around the corner in High Street were the Blue Bell (now the Goat Major), the Cardiff Castle, Griffin and Wheatsheaf. Off Church Street was (and still is) the Old Arcade. Situated in St. Mary Street were the Royal Oak, Black Lion, Blue Anchor, King's Head, and Cardiff Cottage, plus the Lamb and Flag, whilst in Queen Street there were the New Inn, Prince Regent and the Cross Keys, located near the East Gate, and used by the Marquess of Bute for his annual court leet for the Manor of Roath.

However, one of the most interesting, and long-departed public houses in pre-industrial Cardiff was the Kemeys-Tynte Arms, later known as either The Tennis Court or Cefn Mably Arms – on the site of the modern-day Owain Glyndŵr pub – adjacent to St. John's Church, which was frequented by the Bute family, as well as the town's mayor, with a pew and mace-holder still surviving where he sat for the civic services.

The Kemeys-Tynte Arms had originally been the town house of the landed local family based at Cefn Mably whose male heads, at various times, were MPs for Monmouthshire, High Sheriffs of Glamorgan, and governors of Cardiff Castle. During the 18th century a court for real tennis was constructed in the garden of the house. The game was the original racquet sport from which the modern form of tennis evolved, and is known as the sport of kings, largely through its association with Henry VIII who played the game at Hampton Court, as well as its references in Shakespeare's *Henry V*.

It is therefore reasonable to assume that it was here, close to St. John's Church, that the well-to-do young gentlemen of the town assembled during the pre-industrial era to engage in healthy recreation, thereby mimicking the patterns of behaviour they had learnt at public schools in London and south-east England as well as having seen in action at Oxbridge colleges and elsewhere, including the Marylebone Cricket Club at Lord's where a real tennis court also existed.

When the Kemeys-Tynte's townhouse became vacant, the premises were converted into a tavern and, with the tennis court alongside, it would have been a very convivial place for the wealthy youngsters, as well as older gentlemen to congregate. In fact, the sporting young men of the sleepy market town may have been meeting in that vicinity for some time as there are records from the 18th century of damage being caused by ball games in the vicinity of St. John's Church with balls being thrown and hit against the north wall of the church tower.

The Kemeys-Tynte Arms, or The Tennis Court, may have been amongst the first public houses in the market town where gentlemen gathered. Another place may have been the White Lion Inn, known to have been situated since the 1770s on Castle Street, as there are references in the *Cardiff and Merthyr Guardian* newspaper during the 1840s to Cardiff Cricket Club using the premises for pre-season meetings.

By this time, however, the number of public houses had dramatically increased following the opening of the Bute Docks in 1839, with around 30 pubs lining Bute Street alone, including the notorious Custom House Inn, and at the northern end of the street the Golden Cross, opened in 1849 as the Shields and Newcastle Tavern. Just as the number of watering holes increased, so did the number of drinkers, especially dock workers, labourers and sailors, all eager to quench their thirst and to meet friends and mingle with the women of the town.

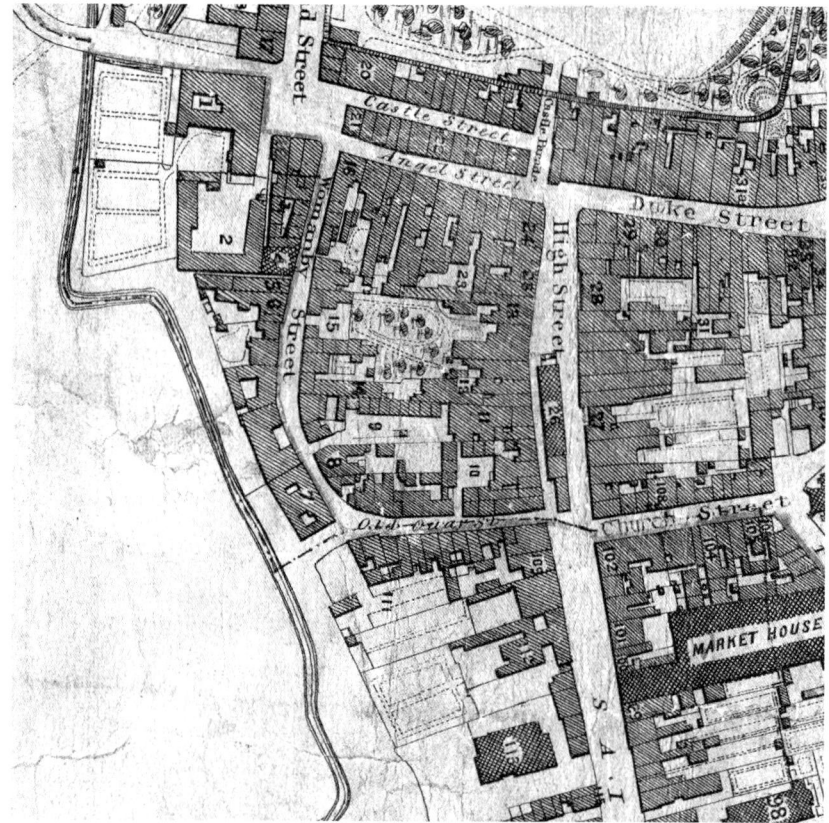

The Cardiff Arms is marked as 'building 1' on O'Rourke's 1849 map of the town.

Many of the city centre taverns therefore got busier following the opening of the docks and the rapid increase in the town's trade. It prompted many of the gentlemen of Cardiff and surrounding area to seek more refined and genteel surroundings elsewhere, especially in the town's hotels, including the Cardiff Arms which had opened in 1792. Indeed, their preference to meet up in a hotel no doubt explains why the next chapter in this story about the evolution of the Cardiff and County Club focuses on the creation of The Royal Hotel in St. Mary Street.

2

Mr. Heard and the Hotel Club

"We have a lot to thank those pioneers of the nineteenth century for in creating this Club and setting the traditions and standards by which we still stand."

Henry Heard was typical of the English-born gentlemen who, during the second half of the 19th century, migrated to live and work with great effect among the professional classes in the thriving town of Cardiff. Born in Devizes in 1832, he had attended Chippenham Grammar School before moving from Wiltshire to south Devon where he commenced his legal studies and training in Exeter before heading to south Wales and working in a practice in Abergavenny. Whilst he was based in Exeter, he met and wooed Mary Ann Helmore, the daughter of a well-to-do gentleman from Heavitree, and on December 29th, 1855 the pair were married.

They spent only a short while in Abergavenny, as in 1857 Henry acquired a senior position in a practice at 24 Trinity Street in Cardiff. His *forté* was property law and he swiftly made his mark by being involved, during 1858, in the creation and promotion of the first shopping arcade in Cardiff. The Royal Arcade, as it became known, linked St. Mary Street with The Hayes, and three years after its opening, a free library and museum was created by voluntary subscription above the entrance in St. Mary Street, with Henry serving as a Vice-President of the new institution. It was the place where famed *Western Mail* cartoonist Joseph Staniforth undertook his studies whilst, during the 1890s, David Morgan's department store was created over, and either side of, the Arcade. It was through his dealings with the creation of the Arcade, as well as the Free Library that Henry made his name and came into contact with the men with whom he would subsequently form the Cardiff and County Club in 1866 at The Royal Hotel.

Within ten years of arriving in Cardiff, Henry and Mary Ann had three sons and two daughters, with the well-to-do pair making their family home

A stamp book dated 1861, in the handwriting of Henry Heard. Did it belong to an 'informal' gentlemen's club who met and wrote letters in one of Cardiff's public houses?

at 27 Charles Street. Besides being highly active in the legal world of south Wales, Henry and his wife threw themselves into the social world, becoming actively involved in cultural and musical events, billed as Cardiff's 'Winter Evening *Conversazione*', held in the ballroom at the Cardiff Arms Hotel. They were also active members of Cardiff's Naturalists Society, a hotbed of intellectual discussion, allowing further opportunities for the Heards to mingle with the great and good of Cardiff society.

Henry also became involved with the Oddfellows' Widows and Orphans Society, and made a number of generous donations towards the establishment of a school in Llandaff for deaf and dumb children. He also joined the Fourth Glamorgan Artillery Volunteers, and during November 1861 was gazetted as a second lieutenant. He became actively involved in local politics and in July 1864 was elected as the member for the North Ward of the Town Council. As a staunch Conservative, he was prominent in local affairs for several years, and during 1876 oversaw Cardiff Corporation's financial relief effort for the Indian Famine.

As far as his practice was concerned, amongst Henry's clients were the Bute Estate as well as Lord Tredegar, who lived at Tredegar Park in Newport and, like the Butes, owned vast swathes of land across the coastal plain as well as in the mineral-rich valleys. Henry oversaw the arrangements for the leasehold development and sale of properties on the Tredegar Estate's land in Splott and Canton, as well as on the Marquess' land in Crockherbtown, Roath and Butetown, plus the disposal of the Tredegar-owned Cae Castell Estate in St. Mellons.

The Butes and Lord Tredegar were just two of the grandees in the social and industrial world of south Wales on whose behalf Henry acted. Within

Cardiff in 1851, showing the site of the railway station and the old course of the Taff.

barely half a dozen years since his arrival in Cardiff, his stock had swiftly risen to the point that, in the latter months of 1863, he become a key member of the limited company set up to create The Royal Hotel – and the first home of the Cardiff and County Club – at the southern end of St. Mary Street.

By this time, a second dock had opened close to the mouth of the Taff, following the completion in 1859 of the Bute East Dock, and by the early 1860s, coal had replaced iron as the leading export, with around two

The Royal Hotel circa 1890.

million tons leaving the port of Cardiff in 1862. Some colliery agents and ironmasters were concerned about the charges levied by the Bute Estate to use their facilities, so a plan was set in motion to open a rival dock at Penarth. There were also concerns in Cardiff about some of the antiquated and inadequate facilities; in many ways, the hotel facilities still resembled those of a sleepy market town, not a vibrant and go-ahead industrial complex. In both respects, there was a need to keep Cardiff as a port, and a visiting place, as attractive as possible, so the creation of the hotel, and indirectly the Club, by the Cardiff Hotel Company Limited was an integral part of the way Cardiff would move forward.

Henry acted as both solicitor for, and the general secretary to, the Company as, in the words of a notice placed in *The Cardiff and Merthyr Guardian* on February 26th, 1864, they:

> 'sought to raise, through share capital, funds for the erection of a first-class family and commercial hotel in Cardiff. The immense increase in trade and population of the town and port has not been accompanied by a corresponding increase of hotel accommodation. The fact that there is, at present, only the same number of hotels in the general business part of the town as existed fifteen years ago, although the population

has more than doubled within that time, is a sufficient guarantee that a new hotel built and arranged on modern principles and managed in a liberal and judicious manner, must prove a remunerative investment.'

The site which the company acquired, through Henry's contacts with the Bute Estate, lay adjacent to the junction between St. Mary Street and the road which became known as Westgate Street, whose creation the Bute Estate had overseen along the eastern border of the Arms Park. The site included a public house called the Stogumber Inn and, together with its adjoining storeroom, the Company planned to demolish the premises and in its place create, in the words of the share prospectus, 'a handsome structure in the Italian style.' The architect and civil engineer, Charles Bernard, won the tender to design the building and during 1864 financial support was duly secured for the scheme.

It was not the only hotel scheme in the southern part of St. Mary Street as, with sizeable financial support from the Great Western Railway, another large and ornate complex was planned on the junction with Station Terrace itself. However, the major difference between what became known as The Great Western Hotel and the Royal Hotel was that Henry Heard's scheme had a Club – 'with private Reading, Dining, Billiard and Smoking Rooms which will communicate with the Hotel' – as an integral part. It is not known who had been the prime mover behind the creation of such a facility but the solicitor and his friends from the legal, political and social world of Cardiff must have been very thankful that the absence of such a Club was, at long last, going to be rectified.

Many prominent members of the business world of Cardiff and south Wales were on the Board of Directors of The Cardiff Hotel Company, including timber merchant and town alderman William Alexander who lived in Park Place. The coal trade was also well represented with Sidney Howard, a leading coal agent based at the Bute Docks, Samuel Nash, a coal importer and shipbroker, Lars Ohlsen, a coal and iron merchant who lived at Llandaff Rise, whilst colliery owners Thomas Heath and James Insole were also on the Board. Insole lived at Ely Court and was something of a country gentleman, as he enjoyed hunting and organised shooting parties from his spacious home in Fairwater, often with his good friend Charles

Richard Evans Spencer – a member of the Board of Directors of the Cardiff Hotel Company.

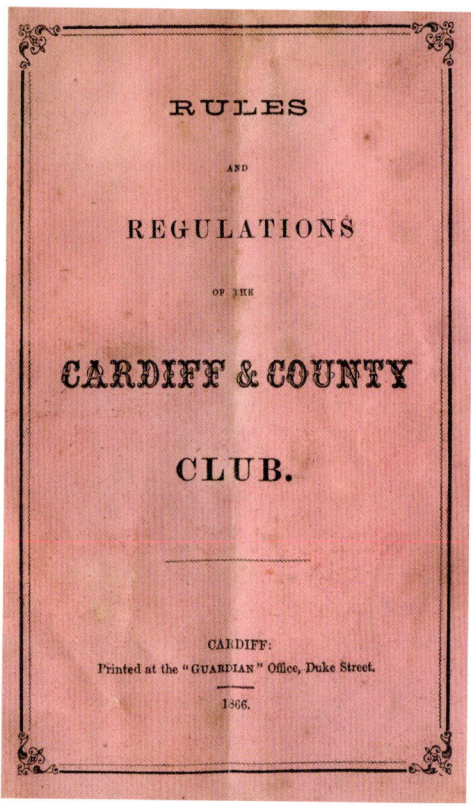

The Club's first rule book, 1886.

David who also acted as a Director of the Hotel Company.

Other members of the Board included prominent figures in the local gentry, including Charles Williams of Roath Court, who in 1872 served as High Sheriff of Glamorgan, and George Fisher, a leading engineer on the Taff Vale Railway, who was a neighbour of Heard, living at 8 Charles Street. Others were friends and acquaintances from the legal world including John Bird of Crockherbtown, and Richard Evans Spencer of Llandaff, a leading cricketer with both the Cardiff club and the Glamorganshire side which had been created during the 1860s in an attempt to form a county side. Also on the Board was George Worthington, his colleague in the club and county cricket team. The presence of these sporting gentlemen proved to be of great importance during the course of the next few years as the Club found new premises and a permanent home adjacent to the recreational heart of the Victorian town.

The Chairman of Cardiff Hotel Company Ltd was Charles Page, a wealthy merchant who lived at Dulwich House in Llandaff. Together with Heard, he oversaw arrangements for the incorporation of the Company on February 15th, 1864. Through their collective energy, drive and diverse network of contacts, they swiftly raised the necessary capital of £15,000 through the issue of 600 shares at £25 each. During May and June 1865 negotiations took place for the appointment of a manager, and after a busy phase of construction work the new hotel was completed by the late summer of 1865. A licence for the sale of alcohol was granted by the Town Council on September 27th, 1865 and, during the autumn, work began on the fitting out of the 50 bedrooms and the 150-seater banqueting room, whilst the rooms where the Club was to be established were also completed.

During December, Henry Heard convened a meeting in the new building at which a resolution was passed that 'a Club be formed in connection with the Town of Cardiff and County of Glamorgan.' Amongst those invited to the meeting were four of the Directors of the Hotel Company – Charles David, Charles Page, Charles Williams and George Worthington – as well as a number of gentlemen who lived in the Crockherbtown area, the prosperous inner suburb north of Newport Road and close to Roath Road (now City Road). They included two solicitors and attorneys – Richard Reece, who was the County coroner, and Richard Wyndham Williams – as well as Dr. Henry Paine, who was the Medical Officer for the Board of Health.

Others present at the foundation meeting of the Cardiff and County Club were some prominent residents from the Vale of Glamorgan including Samuel Gibbon, a master butcher and farmer who lived near Barry; Richard Bassett of Bonvilston House, a magistrate and Deputy Lieutenant of Glamorgan; Lewis Knight Bruce of St. Nicholas Manor House, a magistrate and a member of the family from Aberdare, whose head, Henry Austin Bruce, was a Liberal MP and the first Baron Aberdare; plus Robert Jenner of Wenvoe Castle, whose father had been a High Sheriff of the county and whose family were major landowners in the area.

It was therefore a meeting representative of the great and good from Cardiff and its immediate environs. A further meeting was held in the Town Hall in High Street on January 2nd, 1866, chaired by Robert Jenner, at which the Club was formally constituted together with a committee. They discussed a sliding scale of subscriptions and entrance fees, a membership limit of 300, as well as the various rules and regulations ahead of the formal opening on Monday, April 2nd, 1866.

The prospectus of the Cardiff Hotel Company.

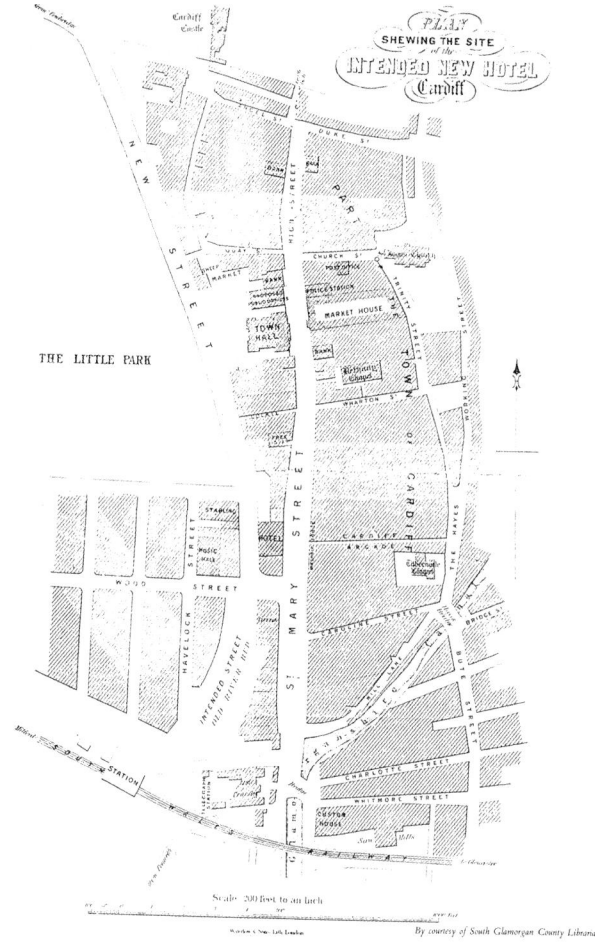

The street plan showing the location of the new hotel.

The opening was accompanied with much pomp and ceremony, with visitors admiring the décor of the new hotel as well as its state-of-the-art innovations, including ventilators to remove gas fumes, press button service bells, and a fireproof staircase at the rear of the kitchen in case of emergencies. The founding members of the Club also much admired the facilities as they were shown around by the hotel's newly-appointed proprietor, Richard Wain.

By this time, 140 gentlemen had joined the Club which hired from the hotel – initially at £250 per annum – four rooms, with a reading room and vestibule on the ground floor plus a billiard, dining and drawing rooms on the first floor. A series of private doors kept the Club separate from the hotel, with the rooms on the upper floor only accessible from the Club's vestibule. Members of the Club were also served food and drink from the hotel at an agreed tariff.

Entrance to the Club's premises was via St. Mary Street, with the hotel having an adjacent foyer and hallway, plus a rear entrance from Park Road (or Westgate Street). Given their juxtaposition, the Club for several years was referred to as The Hotel Club. As members started to use the plush facilities, Henry Heard continued in his role as Honorary Secretary whilst William Bradley Watkins, the Cardiff-born bank manager, borough magistrate and alderman acted as the Club's Treasurer.

Watkins' inaugural accounts showed that the only payments to staff were £52 for the billiard room marker's wages, plus a gratuity of eight guineas to Randall, the waiter employed by the hotel to look after the refreshments ordered by Club members. Within a few years, Randall was living in and employed by the Club at £65 a year, plus £30 for board. A second waiter had also been employed, and the Club's income had grown to £600, from which the rent – which had risen to £300 – was found plus £50 for gas, £40 for washing and £20 for coal.

A noticeboard was also purchased and placed in the corridor leading from the vestibule where telegrams and other messages could be posted, including the afternoon's racing results from the Press Association in London. A wide range of daily, weekly, monthly and quarterly publications was also purchased and, in addition to the local papers, several sporting, social and military titles from London were acquired for members to peruse in the reading room, in which conversation was strictly prohibited. According to the rules, nobody could keep a newspaper for longer than 15 minutes after another member had applied to read it, and anyone removing a newspaper or magazine from the reading room was liable to expulsion.

Members could also be expelled, by the following day, for the non-payment of debts incurred in the card room where whist, piquet, cribbage and ecarte were played. According to the Club rules, stakes were to be no higher than five shillings and no member or visitor was allowed to have debts of more than £3 on any rubber or hand. No games were allowed on Sunday or after 11.45pm on other days, whilst cards were to be used for a maximum of two nights only. The room witnessed some quite heated exchanges, including a pair of blazing rows involving John Sloper. The first occurred in April 1870 when he disputed payment to Randall, the waiter, for a game of whist, and the following year he accused Edwin Knight, a Gloucester-born policeman, of cheating at cards. The pair, plus their playing partners George Fisher and Charles David, were duly asked to prepare statements for the committee, in which Knight was alleged to have winked at David, his partner, who in turn claimed that he did not hear Sloper use the word 'cheat'. Sloper was fully exonerated afterwards.

Fewer disputes were recorded as taking place in the billiard room which was open each day apart from Sunday. Stakes were not to exceed £1 per game of pool or billiards, and anyone cutting the cloth or breaking a cue would incur additional fines or charges. As time went by, there was a steady increase in the number of complaints to Henry Heard about the quality of food supplied by the Royal Hotel, the state of the lavatories, the state of the carpet in the billiard room, and the general condition of Club furniture.

Founder Members

James Ware, a well-to-do ship owner and businessman, was a founder member of the Club in 1866 and typical of the many Englishmen who moved to south Wales to make their fortune during the second half of the 19th century. Born in Bridgwater in 1825, he had initially been the owner and master of a small vessel which plied its trade across the Severn. He duly made his way to the coal metropolis and, by the 1880s, was one of the shipping grandees, residing in a splendid villa in Penarth. Besides being a ship agent, he was also a colliery owner and chairman of the Junction Dry Dock Company, and a Justice of the Peace for Cardiff.

He had arrived at the West Bute Dock during 1850. This was a time, according to the *Cardiff Argus* in September 1888, 'when offices were almost unknown; when commercial transactions were frequently done during a friendly chat with the ship's master on the Pier Head, or what was, perhaps a more convenient shelter, the parlour bar of an hotel; when clerks were few;

A sketch of James Ware published in The Western Mail.

when telegraphic and telephonic communication with the colliery and office had no existence and frequently clerk and office boy were rolled into one.' Sensing a gap in the market, Ware was one of the first shipowners to seek a formal partnership with collieries and he duly became the coal agent to the mines owned by Messrs Sheppard and Evans at Ynyshir, with high grade anthracite being shipped to England and France.

Ware had many good friends of influence in and around the town, especially William Bradley Watkins, who was the first Treasurer of the Club. Ware was also widely known as 'a jolly good fellow' and a man who could be seen daily around the town in a dapper suit with a fresh flower in his button hole. He had very much been in the right place, at the right time, and in 1883 he was one of the generous donors of £1,000 to the fund for the creation of the Cardiff Royal Infirmary, the same amount as donated by the likes of the Marquess of Bute, Lord Tredegar, the Earl of Plymouth, and the Insole family of Ely Court.

Some members also complained about the increasingly impertinent manner of Randall and the lack of service after 11pm.

The wider issue of Club activities after 11pm became something of an annoyance following a visit to The Royal Hotel by Mr. Freeman, the Superintendent of Police. It prompted a sternly-worded letter by Heard to Freeman stating: 'You have said that if the Club is not cleared by 11pm.,

23, Trinity-street, Cardiff,
Dec. 23rd, 1865.

Sir,

The CARDIFF HOTEL COMPANY having provided for the accommodation of a CLUB, at their NEW HOTEL, a Meeting of Gentlemen took place on Saturday last, at which the following Resolutions were passed:—

> "That a Club be formed in connection with the Town of Cardiff and County of Glamorgan."
>
> "That the Gentlemen present at this Meeting be a Committee, for the purpose of forming the Club, with power to add to their number."
>
> "That the Gentlemen named in a List, settled and approved of by this Meeting, be invited by private circular to join the proposed Club, and to attend a Meeting at Cardiff, on Saturday the 30th inst."
>
> "That the circular state the proposed Subscriptions should be as follows:—
>
> Gentlemen originally joining the Club:—

Entrance Fee	2	2	0
Annual Subscription	3	3	0
If resident beyond 8 Miles from Cardiff—Annual Subscription	1	1	0

> Ordinary Members to be admitted by *Ballot*, and residing in or within 8 Miles of Cardiff:—

Entrance Fee	3	3	0
Annual Subscription	3	3	0

> Honorary Members resident beyond 8 Miles from Cardiff:—

Entrance Fee	3	3	0
Annual Subscription	1	1	0

> "That the Gentlemen named in the List, who may be unable to attend the proposed Meeting, be requested to state in writing, to the Hon. Secretary, their willingness or otherwise to join the proposed Club."

In pursuance of these Resolutions, a Meeting will take place at the GRAND JURY ROOM, at the TOWN-HALL, CARDIFF, on SATURDAY, the 30th inst., at half-past 2 p.m.; and should you be unable to attend it, I shall be obliged by your informing me if your name may be placed on the List of Gentlemen originally forming the Club.

I am, Sir,
Your obedient servant,
H. HEARD,
Hon. Sec.

Henry Heard's invitation to prospective members of the Club.

you will go in and clear it yourself. This is a private Club and you have no authority to enter it.' Heard also used his network of contacts to secure further legal advice, and contacted the Home Department in Whitehall to clarify the Club's independent standing.

Heard's exchange with the Police Superintendent gradually became forgotten as more and more members found the facilities at The Royal Hotel fell short of their expectations. Despite this, membership levels were steadily rising and it soon became clear that bigger and better premises were required. As Secretary of the Hotel Company, the complaints about the service, food and accommodation became something of an embarrassment for Henry Heard, especially after he had gone out of his way to ensure the inclusion of a Club in the original designs for the hotel.

After a committee meeting in June 1873, work began on establishing a new site for the Club where larger and more lavish accommodation

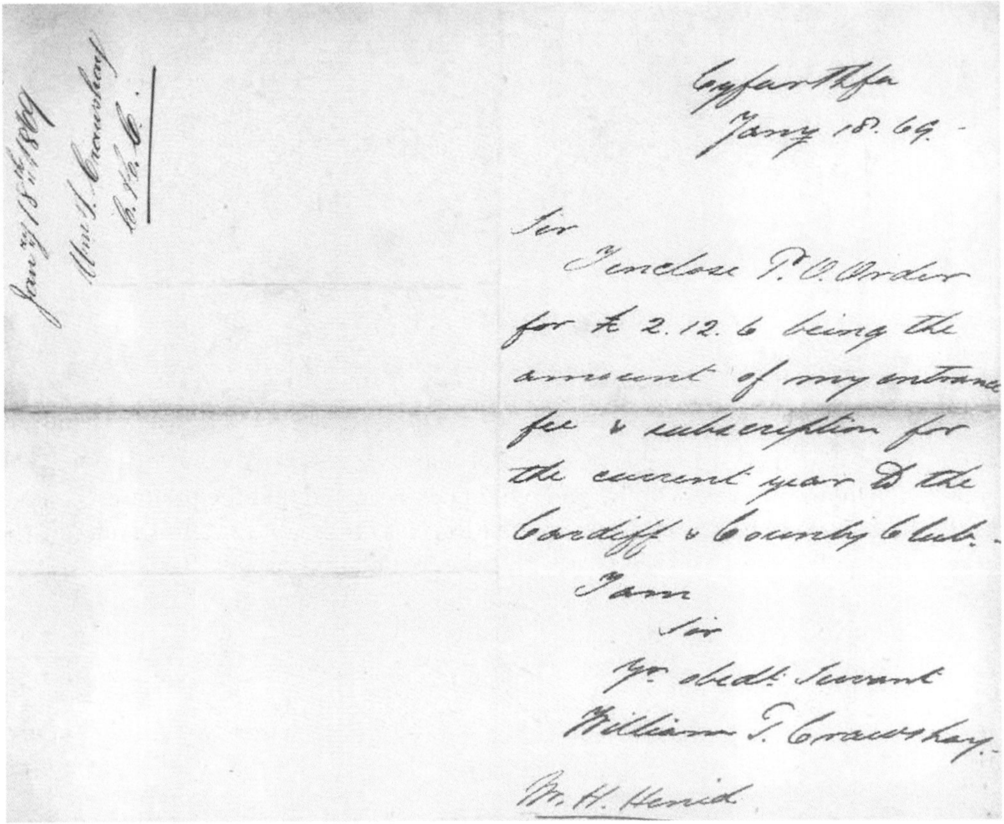

Jany 18th 1869
Wm T. Crawshay
C. & C. C.

Cyfarthfa
Jany 18. 69.

Sir
I enclose P.O. Order for £ 2.12.6 being the amount of my entrance fee & subscription for the current year to the Cardiff & County Club.
I am
Sir
Yr obedt Servant
William T. Crawshay.
Mr H. Heard

A letter, dated 1869, from William Crawshay to Henry Heard.

An early advert for the Royal Hotel.

could be provided. During the next three years, Heard worked with the committee in finding a new home and, when it was agreed that the premises in the hotel should be vacated, he decided that it was an apt time to stand down, and to let a new Secretary oversee the installation of new facilities. He had plenty of other work as well, and overall there was no acrimony over his departure. In mid-November 1876 he formally stood down, with his departure being marked by the presentation of a silver salver and five dessert dishes following a collection among a grateful membership.

Within a couple of years, another Heard was making headlines in the Cardiff-based newspapers as his second son Percy became a highly successful rugby player with Cardiff RFC. During the winter of 1878-79 – his debut season for the rugby club – Percy was the club's leading try scorer with the three-quarter running in 17 tries. Tragically, Percy died whilst on holiday

The Cardiff rugby team which met Newport at Sophia Gardens in the South Wales Challenge Cup on March 8th, 1879. Henry Heard's son, Percy, is pictured standing second from the right.

with his wife and young family in Weston-super-Mare in 1900, aged 39. Given his father's involvement with the Royal Hotel scheme in Cardiff, it was ironic that the Heards should have been staying at The Royal Hotel in the North Somerset resort. Percy had followed his father into law, and had become the senior partner of his father's practice following Henry's death from heart disease at his home, Bute Villa in Newport Road, in November 1894. Percy's obituary in the *Cardiff Times* described his father Henry as 'one of the best known solicitors in south Wales and a man who possessed a very lucrative practice.' Members of the Cardiff and County Club can be forever grateful that Henry Heard found sufficient time to also be the driving force behind the creation of the Club.

3

The Hunting Set

"Horses, hunting and country sports – these were the topics that really interested many early members."

The major task for the Club's committee by the summer of 1873 was to secure rooms of a sufficient size or scope to provide adequate and relaxing accommodation for the members. With vacant premises next door, Heard, in June 1873, invited Thomas Waring, the Borough Engineer, to prepare a report on the feasibility of building a new Clubhouse next to The Royal Hotel.

Waring was the perfect man for the job, with a background in civil engineering and experience as a land agent and an engineering surveyor. He had also practised privately as an architect and had overseen the construction of a vast number of premises throughout Cardiff. Moreover, he was also well known to many of the Club's members as he had played a hand in the

An extract from the Receipts Book – 1867.

> Cardiff 27th March 1869
>
> Dear Sir!
>
> Having been appointed by H. M my Emperor and King as Austrian-Hungarian Consul for Liverpool, I am obliged to leave in a short time Cardiff, and I am sorry I can not therefore continue to be a Member of the Cardiff and the County Club.
>
> Be kind enough to be the interpreter of my friendly feelings to the Members of our Club and believe me my dear Sir
>
> Yours obedient servant
>
> Krapf
> Austrian Hungarian Consul
>
> Henry Heard Esqr.
> Honorary Secretary of the Cardiff and County Club.

An interesting letter of resignation from 1869.

creation of their impressive homes in Crockherbtown, especially in Richmond Road and City Road.

After a favourable report from Waring about the suitability of the site, the Club secured the lease of the premises, with three Trustees – Robert Oliver Jones of Fonmon Castle, Charles Williams of Roath Court and Richard Williams of Parc Thornhill – agreeing to underwrite the acquisition of the lease for 72 years at a cost of £1,200. Waring also drafted some sketches and proposals for the new property, in consultation with Charles David, but it was soon clear to the committee that more money than at first anticipated would be required to cover the cost of both the demolition and the building of the new premises.

Fortunately, seven members were ready and willing to advance the £4,200 required for the next phase of works. Five hundred pounds came from Colonel John Richards Homfray, the son of Jeremiah Homfray and cousin of the cricket-loving George who regularly visited the Club and Royal Hotel, along with many of his sporting pals and other members of the Homfray clan, especially when the Cardiff Races were taking place. Colonel Homfray lived at Penlline Castle, a fortified and castellated Norman mansion near Cowbridge, having purchased the property in 1846. It was here, together with his wife Maria, that he set up home and had a son, Herbert Richards Homfray, who was to follow in the Colonel's footsteps and to become a leading figure in the life of the Club.

Charles and Richard Williams also provided £500, as did Charles David, whilst George Worthington contributed £700. George Williams of Llanrumney Hall put up £500, and £1,000 came from William Sheward Cartwright, a well-known solicitor and colliery owner from Stow Hill in Newport, who was a leading figure in the racing world of south Wales. Besides being a member of the Jockey Club, Cartwright acted as a steward at many steeplechases and point-to-point meetings on both sides of the Severn Estuary, as well as breeding a number of decent racehorses from his farm, including George Frederick, a stallion which in 1874 won the Epsom Derby.

Charles Williams of Roath Court.

The Derby winner who helped to fund the new Clubhouse

George Frederick was the apple of William Cartwright's eye. Trained by Tom Oliver at Wroughton in Wiltshire, the colt was unplaced on his racecourse debut at York, before winning on his next appearance at Doncaster, and twice more at Newmarket during their autumn meetings. He won again at Newmarket the following May before running at Epsom in a 20-strong field on June 3rd, in the 1874 Derby. Skilfully ridden by Harry Custance, George Frederick moved through the field as it rounded Tattenham Corner, before taking the lead on entering the home straight. Cartwright's horse – which had started at odds of 9-1 – pulled away from the others and crossed the line, becoming the first, and so far only, Welsh-bred winner of the world's most famous flat race.

Cartwright named many of the horses he bred at his stud farm – situated between Cardiff and Newport – after members of the Royal Family, with George Frederick being named after the second son of the Prince of Wales. Cartwright named other horses after places in and around Cardiff; 'Fairwater' won the Great Northampton Stakes in 1863, and 'Ely' won the Doncaster Stakes, the Goodwood Gold Cup and the Ascot Gold Cup. 'Caerau' and 'Penarth', though, met with less success.

With plenty of prize money from these racing exploits, as well as the profits from his family's other businesses – plus his fondness for all things relating to Cardiff – Cartwright was readily able to assist his Cardiff-based

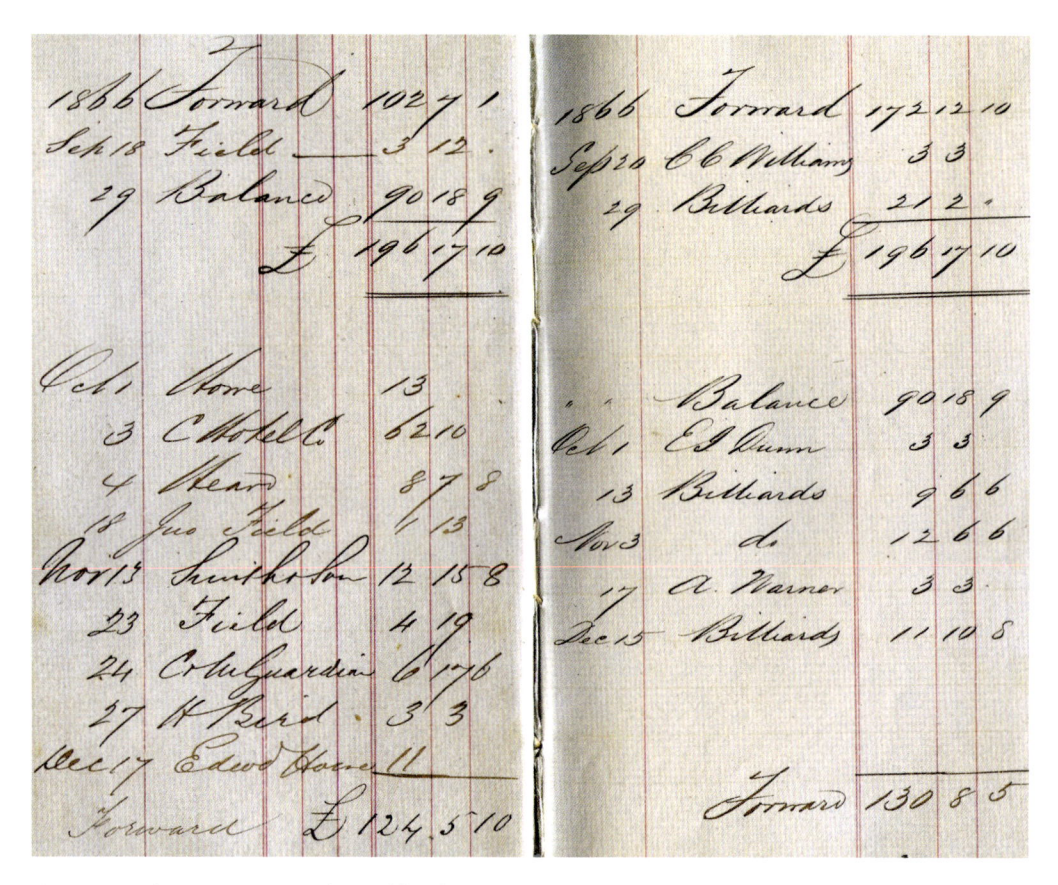

An extract from Henry Heard's cashbook, 1866.

friends in providing almost a quarter of the amount needed for the building work. There were other reasons for his family's benevolence towards the Club, as his son William George, who lived at Springfield in Newport, was also a close friend and business partner of George Worthington, with the pair having major interests in several collieries as well as the Penarth Dock Company.

Like Worthington, he was also a decent cricketer and played many times for the Monmouthshire club, besides forming his own XI which played in exhibition and fund-raising games at the Rodney Parade ground in Newport. He also owned several steeplechasers and shared his father's love of horse racing and point-to-points. With his social and sporting credentials he was a fitting choice in 1883 to be appointed High Sheriff of Monmouthshire. Tragically, he was not to see out his term of office as he was taken ill shortly before Christmas and died early in the New Year at the age of just 49.

An extract from the subscriptions book, 1866.

Cartwright was just one of many members of the Club at this time who had a keen interest in hunting. Indeed, there were so many with an interest in equine matters that, on Friday, April 22nd, 1887, a steeplechase took place, solely for Club members and their horses with a four-mile race taking place between Llanrumney Hall and Castleton. The race was on land owned by Lord Tredegar and Charles Williams of Roath Court, whilst the other arrangements and entries were overseen by Francis Villiers Bruce, the son of founder member Lewis Knight Bruce of the Manor House in St. Nicholas.

Lewis Knight Bruce dined, on average, three times a week at the Club and was a noted figure in the hunting world of south Wales. Eager that his 22-year-old son made a name for himself and followed in his footsteps, Lewis encouraged Francis to organise the steeplechase, which was the first point-to-point in south Wales. He received encouragement from Charles Berkeley, the hard-working Secretary of the Club who himself was a keen

Henry Heard's members list from 1866-67 reads like a 'Who's Who' of south Wales.

horseman and huntsman, besides being a leading figure in the organisation of the Cardiff Horse Show.

Twelve horses were duly entered for the 1887 race, with the riders including Frederick Courtenay Morgan (the son of Lord Tredegar), Hugh Homfray, Charles Williams, and The Mackintosh of Mackintosh, as well as Francis Bruce himself. George Crofts Williams of Llanrumney Hall agreed to host a lavish lunch so, from around 1pm that afternoon, the great and the good of Cardiff society, plus the sporting members of the Club and their ladies gathered at Llanrumney for the sporting entertainment.

After lunch, the gentlemen and ladies walked to a field near St. Mellon's Church where Lord Tredegar, Colonel Godfrey Morgan, was sat astride one of his hunters. The man, who had taken part in the Charge of the Light Brigade at Balaclava in October 1854 and later became MP for Breconshire, sounded his hunting horn as the dozen participants set off on their race. It proved to be keenly-fought as the horses made their way towards a flag placed on part of Duffryn Farm, owned by Richard Stratton at Castleton, and then galloped back towards Llanrumney.

Over the last two fences, Francis Bruce took the lead, with his mount Saucy Boy – owned by Robert Forrest, the land agent for the Earl of Plymouth – crossing the winning line in front of Redwing, ridden by Charles Williams' son Bobby. The event was a huge success, especially as it was the perfect opportunity for the sisters, daughters and wives of Club members to join with their men folk and mingle with other members of local society.

The contest was repeated in April 1888 with ten horses taking part in a slightly shorter race between Duffryn Farm and Marshfield Church. This time the pre-race entertainment took place in the palatial setting of Tredegar House, so it was somewhat fitting that the race should be won by Freddie Morgan on a horse called Rambler. In 1889 the steeplechase was held in March across land near Bonvilston Turnpike, and was overseen by George Thomas of The Heath, Colonel Hobart Tyler, Henry Lewis of Greenmeadow and Samuel Gibbon of Cowbridge. The winner was a horse called Spider, ridden by David Gore Lindsay, the son of its owner Lieutenant-Colonel Henry Gore Lindsay. Lindsay's other son, Morgan, rode Freebooter into third place whilst Edmund David was among those whose horse was unplaced in the 18-runner race.

For the next few years, the race continued to be staged, and well supported, on land in the Vale of Glamorgan. The annual race helped to settle a few debates and discussions as, prior to the steeplechase, much was said within the walls of the Clubhouse about the merits of various steeds and their riders. Even on the day of the race, there was lots of jolly banter and leg-pulling while, after the event, there were the almost inevitable hard luck stories, as well as plenty of other equine-related topics for members to discuss, adding to the camaraderie and atmosphere of their new premises in St. Mary Street.

4
On the Move

"The Club has always prided itself on good food, excellent wine and a friendly atmosphere."

It wasn't really much of a move as the Club, at first, went from 66 St. Mary Street to next door at number 67 but, at least, the new premises gave the Cardiff and County Club a much clearer identity and independent presence. Although there were linking doors with The Royal Hotel, the new building had its own manager and, with kitchens and staff in the basement, the catering demands of members could be dealt with – and promptly – by their own people, rather than being reliant on those from the hotel which had led to delays at busy times.

The creation of the new premises began during May and June 1874 when Henry Heard sent tender letters to approximately 40 architects throughout the UK with the invitation containing the brief drawn up by Thomas Waring that the new premises should comprise:

- a basement with a kitchen, scullery, larder, wine and coal cellar, plus a servant's hall;
- on the ground floor, a dining room, reading room, bar and waiters' room, lavatories and bathroom, plus a lobby, vestibule and private doors into the hotel,
- a first floor with a card room, visitors' writing room, billiard room, dressing room and toilets; and
- a second floor with three or four bedrooms for staff.

Advice was also sent that the price for the new premises should be 'not more than £3,000 to £4,000 at the most'. Evidence of the great interest shown in the plans can be gauged from the lengthy report in the *Western Mail* of Monday, July 13th, 1874, containing details of the various submissions which Heard received, plus comments from the committee that they hoped

Private.

CARDIFF,

23RD JANUARY, 1866.

SIR,

The Cardiff Hotel Company having provided, in their new building, Reading, Dining, Billiard, and Card Rooms for the accommodation of a Club, a meeting of gentlemen took place in the Town Hall, Cardiff, on the 2nd instant, (R. F. L. Jenner, Esq., in the chair), when a Club was formed, under the name of "THE CARDIFF AND COUNTY CLUB."

The following gentlemen were appointed a Committee to draw up Rules and Regulations for the Management of the Club, and to submit the same for the approval of the Members at a meeting to be called for the purpose, viz. :—

RICHARD BASSETT, ESQ.	C. W. DAVID, ESQ.
LEWIS K. BRUCE, ESQ.	CHARLES H. PAGE, ESQ.
J. SAMUEL GIBBON, ESQ.	H. J. PAINE, ESQ.
R. F. L. JENNER, ESQ.	R. L. REECE, ESQ.
CAPT. PALMER.	GEO. WORTHINGTON, ESQ.
CHARLES H. WILLIAMS, ESQ.	RICHARD W. WILLIAMS, ESQ.

The Committee were empowered to arrange with the Hotel Company for the rental of Club premises, either furnished or unfurnished, upon such terms as they might think fit.

The proposed Entrance Fees and Subscriptions to the Club are as under :—

Gentlemen originally *joining the Club* :—

	£	s.	d.
Entrance Fee	2	2	0
Annual Subscription	3	3	0

If resident 5 miles from, and within 20 miles of, Cardiff :—

	£	s.	d.
Annual Subscription	2	2	0

If resident beyond 20 miles from Cardiff :—

	£	s.	d.
Annual Subscription	1	1	0

Members (to be admitted by ballot), residing in, or within 5 miles of Cardiff :—

	£	s.	d.
Entrance Fee	5	5	0
Annual Subscription	3	3	0

Members (to be admitted by ballot) residing 5 miles from, and within 20 miles of Cardiff :—

	£	s.	d.
Entrance Fee	3	3	0
Annual Subscription	2	2	0

Members (to be admitted by ballot) residing beyond 20 miles from Cardiff :—

	£	s.	d.
Entrance Fee	2	2	0
Annual Subscription	1	1	0

At the Meeting which took place on the 2nd inst., it was resolved that the gentlemen who had expressed to the Secretary their intention of joining the Club, or might do so on or before the 31st of January instant, should be considered as *original* members; and after that date they should only be admitted by ballot.

~~Should you be desirous of joining the Club, I shall be obliged by your writing me a note to that effect by the 31st instant, when the list of original members will be clos~~ed.

On the other side I send you a List of the gentlemen who have already consented to join.

I am, Sir,

Your obedient servant,

H. HEARD,

HON. SEC.

The Club's initial subscription rates.

11th April 1870

Dear Sir,

I am requested by the Committee to direct your attention to the rule relative to the introduction of visitors to the club, and to state that your introduction of the French Consul on the 5th inst. is irregular.

Yours faithfully

B. Heard

Hon: Sec:

E. J. Knight Esq
Charles St.
Cardiff.

The Club's rule, concerning visitors, being asserted in 1870.

the design would prove to be 'an ornament to Cardiff, and will be a step in the direction of providing the town with buildings worthy of its increasing progress.' The tender was eventually won by Wilson, Willcox and Wilson of Bath. Jacob Biggs, the contractor based at Cambria Villa in Roath, secured the tender for the building work and on October 14th, 1874 a contract was drawn up with him agreeing to complete the work by September 1st, 1876.

A slight glitch occurred in January 1875 when the Town Surveyor stopped work on the new building because a bay window projected three inches too far into St. Mary Street, and beyond the line of the adjoining hotel. The plans had been for the new building to 'harmonise with the Royal Hotel' so after some sharp correspondence between all parties, Biggs was instructed to move the window back by three inches.

Later in the year, he encountered further problems because the plot adjoining Westgate Street comprised around 15 feet of mud rather than solid bedrock; the mud had been placed into the remains of the old river bed following the diversion of the Taff by Brunel and his colleagues. Work was therefore halted until a secure base level had been found and a thick layer of concrete had filled the gap up to ground level.

Fortunately, Heard and the committee met with fewer difficulties securing tenders for the fittings, with the dining and reading rooms being furnished in dark oak and mahogany. The dining room had the best quality Brussels carpet, whilst plain blinds were installed in the windows looking out into

St. Mary Street, with venetian blinds in the rooms at the rear. Other high quality fittings and furniture were secured to cater for members, who now numbered over 250, and provision was made to have sufficient equipment for a further increase in membership closer to the limit of 300.

By early December 1875, work had progressed sufficiently for an advert to be placed in *The Cardiff Times*, the *Telegraph*, and in the *Army and Navy Gazette* for a Messman, preferably married and without children, whose wife was a good cook, and who would undertake the general management and provide provisions, besides looking after a team of kitchen maids, housemaids and servants.

This marked a major move forward in the Club's evolution, as no longer would members have to rely on refreshment from the hotel. Given the many complaints which Heard had received about the catering services provided by the hotel, it was imperative that an experienced couple were employed. Applications came from staff attached to Princes Club in Manchester, as well as the Bath and County Club in Somerset, but the eventual choice was Charles and Elizabeth Chalk, who were in charge of the Officers' Mess at Cheriton Camp in Kent. Born in Newport on the Isle of Wight, Chalk had plenty of catering experience, and he and his wife were in post when the Club was formally opened to members on Monday, March 7th, 1876.

Shortly afterwards, John Morgan, a solicitor who lived at Claremont House in Tredegarville, took over the Secretary's duties at an agreed sum of £60 a year and, for the next few years, he and Chalk efficiently ran the Club. Evidence of their success can be gauged by the steady influx of applications for membership, whilst the Chalks' services were outsourced for other civic events, including the Cardiff and South Wales Horse Show, held at Sophia Gardens, plus the Christmas Dinner organised for the poor of the parish of St. Mary's and held in the Bute Terrace schoolrooms. They also worked at various functions at the Town Hall, including a Bachelors' Ball, where the sons of Club members mixed with the daughters of members and other ladies.

Indeed, the Assembly Rooms of the Hall were also used for the Annual Ball which the Club's committee instigated around this time, giving the gentlemen of the Club a chance to dance with their wives and daughters. One of the most active supporters of this venture was James Howell, the owner of the famous department store in St. Mary Street which became known as 'The Harrods of south Wales'. Born in Fishguard in 1835, Howell was a farmer's son who had briefly worked in London before moving to Cardiff around 1865 where he set up his first shop in The Hayes, before opening a fabric and furniture store at 13 St. Mary Street in 1867. A carpet and millinery premises soon followed and, within a few years, it was *the* store to visit for his friends and fellow members of the Club. His flourishing business allowed him to build a substantial home in The Walk, Tredegarville, as well as a lavish townhouse known as The Grove in Richmond Road. It was here that James, his wife Frances and their 14 children lived during the late Victorian era.

James Howell played a prominent role in the social life of the Club, overseeing the arrangements for the Annual Ball in the Assembly Rooms, and covering the costs of the rooms' decoration. It was a most popular event and reflected his meticulous attention to detail that had been one of the hallmarks of the success of his popular department store. He died in May 1909 having enjoyed a hugely successful business career. A reflection of his high standing within Cardiff society, as well as the grandeur of his family's home, was that The Grove was subsequently sold to the City Council in 1912 and became The Mansion House.

Another key figure in the organisation of the Club's Annual Ball during the 1870s and 1880s was the Hon. Frederick Courtenay Morgan, the third son of Lord Tredegar, and a resident of Ruperra Castle in Lower Machen. Educated at Winchester, Freddie had followed his brothers into a military career, serving in the Army, before entering local politics and becoming an MP for Monmouthshire. Like many of the members of the Cardiff and County Club, Freddie enjoyed hunting and had been a regular participant in the Club's annual point-to-point, whilst in his youth he had played cricket for both Monmouthshire and the South Wales CC. Perhaps through the influence of Morgan and other Club members, the South Wales CC – which had been inaugurated in 1859 by George Homfray – held their Annual General Meetings in the Royal Hotel during the week of the annual Cardiff Races at Ely.

The Annual Balls, and the improved catering at the Club, saw the standing of Charles Chalk and his assistants rise significantly, and it was fitting that Chalk and his team were invited to organise a lavish banquet held in Pontypridd on August 6th, 1877 when the Lord Mayor of London, Sir Thomas Scambler Owden, visited the town to distribute money which had been raised to help the families of those who had been affected by the Tynewydd Colliery Disaster in Porth. Four months before, a series of workings had been flooded by water from abandoned shafts of the Cymmer Old Colliery, although it was fortunate that the water gushed through the mine shortly after the afternoon shift had ended. Five men were killed, but a further 14 colliers were trapped. Four were rescued within 18 hours, but it took nine days before the others were freed, with the plight of the trapped men and the bereaved families touching the hearts of the nation.

Owden travelled from London to Pontypridd to hand out relief money to the widows and families of those who died in the mine, before a grand luncheon was held in the town, with Chalk overseeing the arrangements for the 350 guests. The meal was paid for, and organised, by one of the founder members of the Cardiff and County Club, the Reverend David Watkin Williams who lived at Fairfield near Trefforest and, at the time, a man regarded as one of the most colourful figures in the area.

Educated at Trinity College, Cambridge, Williams served 15 years as curate of Ystradfodwg but, on the death of his father in 1857, he inherited the sizeable estate which included land in the Taff Valley and in the Vale of Glamorgan. He duly relinquished his clerical duties and became more akin to a country squire. Williams was a familiar face in Pontypridd, and sat on the local Magistrates Board, and was equally well-known in Cardiff, where he frequently met friends at the Club and fellow founder members, talking for hours about events on the hunting field.

A Vanity Fair cartoon of Frederick Courtenay Morgan, drawn by Leslie Ward and published on February 11th, 1893.

His generosity in hosting the grand lunch was a gesture typical of his benevolent attitude towards the people of the Pontypridd area and, when Lord Aberdare spoke at the end of the meal, he paid a warm tribute to Williams' kindness and raised a toast to thank him for his efforts in the distribution of the money to the bereaved. Indeed, it had been to Williams that Sir Thomas had first spoken when seeking help as to who should receive the monies, with Williams providing an initial list of names.

At times, Williams cut quite an eccentric figure with contemporaries describing him as 'very leisurely, tall with somewhat drooping shoulders, a square felt hat, a very open collar and a long black tail coat with white patterned trousers.' In 1868 he 'adopted' the eight-year-old son of his sister, with the young lad called Theophilus living with Williams, his housekeeper and servants at both Fairfield and at Verlands, one of Williams' impressive new properties in Cowbridge.

From afar, it seemed quite an unusual set-up with the young adolescent living with his landed uncle, who remained a bachelor until the age of 63. By the time of his uncle's marriage, Theophilus was 20 and only slightly younger than the bride, Mary Mostyn Renwick, a lady from Bala in north Wales who duly became his adopted mother. Williams died in 1891, with Theophilus inheriting great wealth and rising to a similar high status in Glamorgan, reflecting the series of good business decisions and property acquisitions which his adopted father had made.

One of the many fine decisions which Williams had taken had been to hire Charles Chalk as *maitre d'hôtel* to oversee the grand luncheon in Pontypridd in 1877 when the Lord Mayor of London visited the area. It gave great publicity for the Club to have some of the big-wigs at the luncheon and, judging by the fine comments paid to Williams about the quality of the fare, reflected very well on both Chalk and the Club itself. It was perhaps no coincidence that, shortly afterwards, Chalk's catering interests further diversified when he was also installed, in 1878, as manager of the Cardiff Racquets and Fives Club.

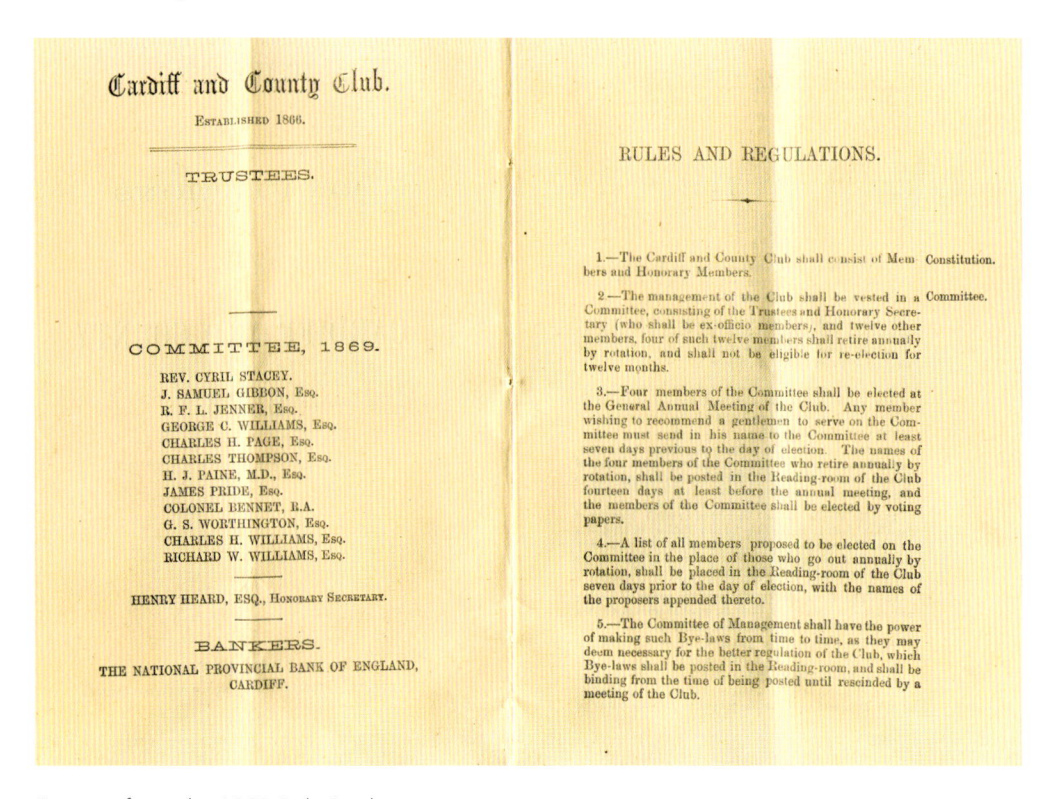

Extracts from the 1869 Rule Book.

The Racquets and Fives Club had been built on Westgate Street, adjacent to the Arms Park, where plenty of sporting activity took place throughout the year. Cardiff Rugby Football Club had come into being in September 1876, and from 1867 Cardiff Cricket Club had its own private ground in the north-western corner of the park, with other local teams using the rest of the space in the park, which extended south towards Park Street.

Many of the Club's younger members took part in these sporting contests, although the physical, hurly-burly of rugby was not to everyone's liking. Many sought less violent and more genteel recreation, so an approach was made to the Bute Estate to build a complex where fives and racquets could be played. Both were staple forms of recreation in the English public schools and creating facilities so that these games could be played represented the influence of the immigrants to Cardiff on the development of recreational facilities within the coal metropolis. In many ways it was throwback to the creation of Cardiff's first sporting arena – the real tennis court, adjacent to the Kemeys-Tynte Arms in Church Street.

An advert for the Cardiff Arms, with the final comments showing how proximity to the recreation grounds was an important selling point.

5

Good Old Frank

"The close link with the Bute Estate has been so important for both the Club and Cardiff itself."

The Third Marquess of Bute was a keen supporter of recreational activity, his interests having been sparked whilst a student at Harrow and Christ College, Oxford, as well as during the late 1850s and early 1860s when he was privately tutored by Frank Stacey, the son of Rev. Thomas Stacey, the vicar of St. John's and subsequently Precentor at Llandaff Cathedral. With Frank having himself attended Eton before reading law at King's College, Cambridge, he was regarded by leading members of the Bute household as a most suitable person to accompany and tutor John Patrick Crichton-Stuart.

Stacey was also one of the finest young cricketers in south Wales and, after appearing for Cambridge University in 1853, he played for a number of gentlemen's XIs in England, as well as appearing for local Welsh teams in games at Neath and Llanelli against wandering English XIs. He also played with distinction in the exhibition matches on the Arms Park in 1855 and 1857, and appeared for the Glamorganshire side that was formed ten years later in 1869. By this time, Frank had established himself as one of the leading barristers in the Cardiff area and lived at Llandough Castle – a fortified manor house with castellated effects situated to the south-west of the town and on a high ridge overlooking the coal metropolis.

His friendship with the Marquess had survived a quite turbulent period in the life of the teenager, following the death on December 28th, 1859 of Lady Sophia Bute from Bright's disease (a form of kidney failure). The pair also spent several quite eventful months in south Wales and Scotland as a row developed over the youngster's custody. Early in the New Year of 1860 three haughty dowagers were appointed as trustees to look after the teenager and for Frank, in particular, life with the Marquess and his entourage of grand dames swiftly became markedly different to what he had experienced as an undergraduate at Cambridge, and as a young gentleman in the town of Cardiff, as well as on the cricket fields of south Wales.

In between lessons, Stacey filled his time with shooting and fishing on the Scottish estates, and during the evenings he kept himself in good spirits, quite literally, by imbibing. On occasions, he also shared a drink with his 13-year-old charge and, as their friendship grew, the Marquess heard more and more tales about cricket, Cambridge, and the good life of Cardiff. These turned the head of the adolescent Marquess who was now being fussed over by quite an austere collection of ladies. The pictures that Frank drew were a million miles away from the sheltered existence the Marquess had hitherto known and it was not long before Stacey suggested to his charge that he should be entrusted to look after his affairs.

This soon led to a spat with Lady Elizabeth Moore, who was one of the trustees, which left the spinster astonished and dissatisfied about Stacey's role. Things came to a head when the puritanical Lady Elizabeth discovered that Stacey had smuggled brandy into Dumfries House and, in a letter to the other trustees she alleged that 'Stacey intended to introduce him to vice, and gradually to have complete power over the unfortunate child'.

The upshot was that Lady Elizabeth suggested that, rather than having a tutor, the teenager should prepare for a place at Harrow by attending the Scottish Episcopal boarding school at Glenalmond near Perth in Scotland. The Marquess disagreed vehemently but shortly afterwards, he was taken away by Lady Elizabeth to the George Hotel in Glasgow. The following day, Frank Stacey turned up and yet another heated argument took place with Lady Elizabeth. She subsequently described Stacey's language as 'maniacal', and his behaviour as 'showing the greatest rudeness and violence', and shortly afterwards, Stacey was relieved of his duties as the Marquess' tutor and, as a custody battle ensued, he returned to south Wales.

Stacey remained in close touch with the Marquess who, in spring 1862 started

The gentlemen of Cardiff Cricket Club, in 1875. Many were members of the Cardiff and County Club, and are seen standing in front of the pavilion financed by the Marquess of Bute.

at Harrow. By this time, a court hearing had also taken place, with Lord Bute being placed under the charge of the Earl of Galloway. Through his encouragement, and that of Stacey, the Marquess learnt the rudiments of cricket at Harrow, before securing a place at Christ College Oxford. It was through this friendship with Stacey that the Marquess became more actively involved with the affairs of Cardiff Cricket Club. After a little bit of encouragement from his irascible friend and former tutor, the Marquess readily agreed in 1865 to become the patron of the town club and in 1867 financed the construction of a pavilion in the northern part of the Arms Park, solely for the use of the town's cricket club, with Stacey also overseeing the arrangements for the creation of the club's exclusive ground.

Frank's brother Cyril, who had succeeded his father as the curate of St. John's, had been a founder member of the Cardiff and County Club before joining the Club's committee. Within a short time, Frank also joined the Club and soon enjoyed everything that was on offer in the premises in St. Mary Street. Indeed, during the summer afternoons, he could relax or dine in the Club, before using the rear entrance into Westgate Street and calling into the Arms Park to watch some cricket.

Like others at the Club, Frank Stacey became irritated by the loud and aggressive manner of a minority of members, especially in the card room and, in 1868, Frank was also involved in an incident involving the French Vice Consul, Count de Chappedelaine. The Frenchman complained to the committee that Frank had spoken in an offensive manner but, after a short

A view from 1870, looking south with Westgate Street and Womanby Street (in the foreground) and Temperance Town (beyond).

enquiry, the Count was instructed to write a letter of apology both to the Cardiff barrister and to the Club's committee.

There were other reasons for the promotion of sport on the Arms Park, especially by the many young Conservatives who formed a goodly proportion of both the membership of the Club as well as the town's cricket club. At the time, though, the town council was dominated by Liberals who believed that the Bute Estate should donate land to the town, especially for a new cattle market. The Arms Park, and its position next to the river, made it an ideal location, and in 1866 a delegation from the town council approached the Marquess to donate the land for the badly needed market. He rejected their suggestion, but in subsequent years the Liberals continued to lobby for the park to be used in other ways.

There was therefore more than a whiff of politics over the decision-making as, during the 1870s and into the 1880s, the Bute Estate continued to assist and promote amateur recreation on the Arms Park. During 1881 the entire area was re-levelled and re-turfed and, following an application from the go-ahead members of Cardiff Rugby Football Club, a grandstand with 300 seats was built at the western, or Canton end, of the rugby pitch.

The entrance to the second clubhouse, as photographed in 1990, with the entrance to the original Clubhouse on the left.

In 1866 the Bute Estate had agreed to the laying out of an area adjacent to Westgate Street for quoits – a game where players threw metal, rubber or wooden rings to land over a small metal post. A town club had been formed in 1879 and, after initially playing on land in Charles Street, they moved in 1883 to the Arms Park where a strip of land had been provided running south from opposite Wharton Street where ten rinks had been created ranging in length from 15 (for ladies) to 21 yards (for men). Within a couple of years, the Quoits Club played fixtures against teams from Merthyr and Bridgend whilst after the contests, the players dined in the adjoining Racquets and Fives Club.

The creation of the Racquets and Fives Club, built in 1878 in Westgate Street, had also been undertaken by the Bute Estate, with the encouragement of influential members of the Cardiff and County Club, of which the Marquess had also been a founder member. The friendship between the Marquess and Frank Stacey – who by now was on the management committee of the Club – and his association with many members was instrumental in the decision to hire a pair of leading local architects – John Prichard and George Robinson, who himself was a Club member – to design, at a cost of £2,300, a covered racquets and fives court.

Around this time the Bute Estate was also overseeing other developments in Westgate Street, especially in its northern section, including demolition of the Cardiff Arms Hotel in 1878. Other street improvement schemes had taken place near the old West Gate into the town, as the groundwork was laid for the creation of impressive and grandiose properties on the Bute-owned land.

The siting of the Racquets and Fives Club, midway along Westgate Street, and opposite the rear of the Town Hall (and what is now Guildhall Place) rather than further south towards Park Street, was part of a series of development works which were taking place in the area. During 1876 Edwin Seward, one of Cardiff's most active young architects, oversaw the extension to the Town Hall, and Cardiff Hippodrome was opened next to the junction of Westgate Street and Park Street, in the south-eastern corner of the Arms Park.

This was clearly an area on the up, and one which the Butes wanted to develop tastefully for their Conservative and sporting friends. Indeed, their grand aspirations were evident in the lavish design by Robinson and Prichard for the Racquets and Fives Club, with red-bricks, so favoured by the Marquess in other parts of the town, and Gothic-style architecture. The hiring of Prichard also showed that the Butes had visions of the new premises being a catalyst for further work as Prichard had become renowned for his outstanding designs of many religious buildings in and around the town

> **George Worthington – the man who dined and died at the Club**
>
> In a bizarre way, it was fitting that George Worthington should take his last mortal breath in the dining room of the Club in which he had taken such an interest, and for so many years had been a stalwart supporter. He had helped to stump up some of the cash for the new premises in St. Mary Street, and had chivvied several of his sporting and business friends to follow suit and fund the initial construction costs.
>
> Worthington certainly liked what was created and the bachelor, after selling off his mining interests in the Gelligaer area, moved to live in Cardiff and based himself permanently in rooms at the Royal Hotel. Each evening he would dine in the Club, and it was there on the evening of June 30th, 1881 that he passed away.
>
> He had been dining with a small group of friends when, around 8pm at the end of their meal, he collapsed and gasped for breath. Edmund Reece, the Cardiff coroner and a prominent solicitor, was dining at a nearby table and quickly tried to help his friend. Frederick Granger, the Bristol-born medic, who was one of the town's foremost doctors, then arrived in the room and attempted heart massage.
>
> Sadly, within a short time, it was clear that the assistance was to no avail, and Worthington was pronounced dead. He had been suffering from heart palpitations for several weeks and, earlier that morning, he had called a doctor to his rooms in the Royal Hotel. The medication the doctor gave eased his problems for a few hours and, feeling better, he was able to walk through to the Club for his usual dinner – tragically, it was the last thing he ever did, and the massive attendance at his funeral bore testament to his wide circle of friends in business, sport and society in south Wales.

including St. John's Church in Canton, St. Margaret's Church in Roath, the South Tower of Llandaff Cathedral, and both St. Michael's College and The Old Registry in Llandaff.

The new Racquets and Fives Club soon proved very popular and, within 18 months, a series of tennis courts had been laid out at the rear of the Club, which duly became the home of Cardiff Lawn Tennis Club. Though these developments adjacent to Westgate Street were good news for the Bute Estate, and the sporting members of the Club, they were also to mark the end of the association between the Club and the Chalks. Charles Chalk's appointment in 1880 as the manager of the Racquets and Fives Club was further evidence of the close involvement of the Club in these developments. But, with plenty of bookings and refreshments required by those playing either indoors or outside, Chalk and his staff were kept very busy.

No letters or other records survive to show any disputes with the Chalks, but John Morgan, the Club's Secretary, continued to receive complaints from members about either the accommodation or the catering. By 1883 Morgan had been succeeded as Secretary by William Rose Harvey, a former neighbour of Henry Heard in Charles Street, and a man who had spent more than 30 years in the business world of Cardiff after moving to the town from his native Dunster in Somerset. At various times he had been a governor of the Cardiff Savings Bank and treasurer of Cardiff Horse Show, besides being secretary of the committee that oversaw the staging of horse races at Ely Racecourse to the west of the town.

Harvey was a busy man and, perhaps reflecting the level of disquiet within the Club, he resigned as Secretary in August 1884, before being persuaded to resume his duties. Perhaps he had been frustrated by the

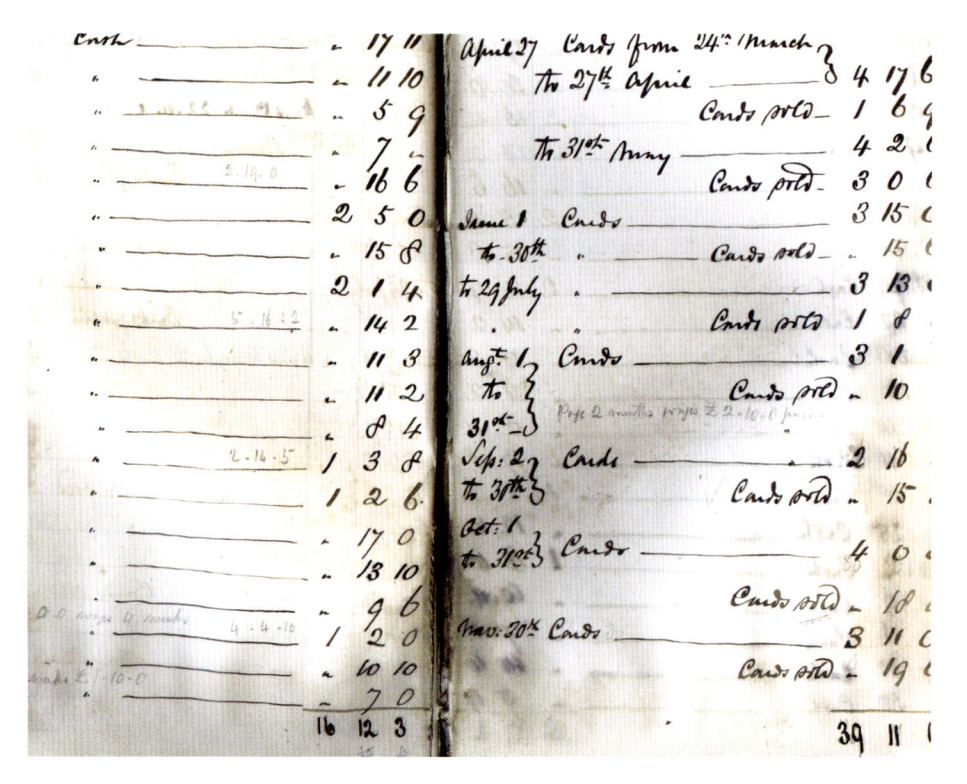

An extract from the 1871 Receipts Book.

A postcard, dated 1909, showing St. Mary Street and The Royal Hotel.

regular ear-bashing from the members, or a volley of letters of complaint. For whatever reasons, Harvey believed that, unlike in the past when Morgan turned a blind eye to the situation, it was time for something to be done, especially as many members believed that Chalk was spending too much time on other ventures and not focusing on the demands of the Club and its members.

Early in 1885, the Club and Chalk parted ways. The quality of his fare and his reputation meant that he was not short of offers, and in March 1885 he acquired the licence of The Borough Arms in St. Mary Street. But, within a couple of weeks of this good news, Chalk was taken severely ill with dropsy and shortly afterwards, he suffered what proved to be a fatal heart attack. It was a tragic end to his time in Cardiff but, through his efforts, and those of his wife and their kitchen staff, they had established a reputation for fine dining in the Club's new premises in St. Mary Street.

Further economic growth, the geographical expansion of Cardiff's boundaries to include Roath, Splott, Cathays, Canton and Grangetown, and the almost tidal flow of more professional people into the town, were reasons why the Club's membership continued to be in a very healthy state and within half a dozen years, the decision was taken to move the Club to larger and even grander premises adjoining the Arms Park.

6

A Permanent Home in Westgate Street

"The design of the Clubhouse echoes the character of the club – respectability, flamboyance, stability and power."

The 1880s was a pivotal decade in the history of Cardiff, and the development of Wales as a whole. Indeed, this decade saw a tsunami-like wave of civic, cultural and sporting nationalism flow across many aspects of life in Wales. In 1881 the Welsh Rugby Union was founded at a meeting at The Castle Hotel in Neath and, in February that year, a Welsh XV met England in an international match, with the inaugural contest taking place at Blackheath in south-east London. Two years later, the National Eisteddfod of Wales was staged in Cardiff for the first time, whilst 1883 also saw the foundation of the University College of South Wales and Monmouthshire.

The new university, which ten years later became a founding member of the University of Wales, was based in Cathays Park with its impressive white-stone buildings being laid out on land generously provided by the Third Marquess of Bute. He had also contributed £10,000 out of the £50,000 needed to create this new seat of learning, but had declined an approach to purchase the Arms Park as a site for the new university. Instead, the Estate Trustees said, 'Lord Bute is not disposed to give the Arms Park as a site for the new University College as he desired to reserve it as an open space for recreational purposes.' The town's sporting population – and millions of Welshmen and women in subsequent years – could thank the Marquess for his decision to preserve the Arms Park site and look instead to the area north of the town centre.

How different it might have been on April 12th, 1884 when the Welsh rugby team ran out for the first time on to the rugby field at the Arms Park to meet Ireland. Having previously staged home games at Newport and Swansea, the Welsh rugby officials had chosen to play for the first time at the home of Cardiff RFC, and 4,000 people thronged into the ground on a chilly and overcast spring day. To their delight, the fine running play of the Welsh backs helped the home team win the first of many internationals at the Arms Park. After the game, many members of the Cardiff and County Club who had attended the match gathered in their premises in St. Mary Street – like so many others in the taverns and pubs in the Taff-side town – to celebrate the achievements of the men in red jerseys.

It was the dawn of another chapter in the sporting history of Cardiff, as well as a further stimulus to the close relationship between the Club and the Arms Park. Within the oak- and mahogany-panelled walls of the Club, several toasts were drunk to the stars of the Welsh team, as well as to the men like Frank Stacey who had guided the Marquess of Bute's hand in the creation of the sporting facilities on the park. The barrister was a Bute man through and through, having married the niece of Mr. Tyndall-Bruce, a trustee of the Bute Estate, whilst his uncle, Edward Priest Richards was the trusted legal advisor to the Marquess. Richards bequeathed to Frank and his brother Cyril, land in Roath where, with the guidance of Thomas Waring, several areas of housing were developed. Frank was also a director of the Taff Vale Railway and the Cardiff Gas Company, besides being a governor of Howell's School and president of the Glamorgan Agricultural Society.

He was also an influential voice within the South Wales Cricket Club which had ambitions of forming a fully representative cricket team to represent Glamorgan. These dreams were shared by many of Frank's friends within Cardiff CC, including their captain and secretary, as well as Club member,

A carde-de-visite from The Angel Hotel, showing the Arms Park to the right.

John Price Jones. He was a prominent architect who had designed several buildings in the town and in the emerging suburb of Canton, besides playing a key role in the creation of several arcades within the business district. Cricket was his passion, and he had been one of the strongest advocates within the South Wales Cricket Club for the creation of a properly constituted county club rather than a gentlemen's team which undertook an annual tour to London.

Indeed, it was Jones who proposed the dissolution of the South Wales CC at their annual meeting in December 1886 at The Angel Hotel. Eighteen months later, having garnered the support of Sir John Talbot Dillwyn Llewelyn, the squire of Penllergaer, Jones convened a meeting at the same hotel, at which Glamorgan County Cricket Club came into being. The meeting in July 1888 also saw the Marquess of Bute elected as the club's president, and the following June several thousand people were present at Cardiff CC's ground as Glamorgan played their inaugural fixture against Warwickshire.

John Price Jones – as seen in a sketch by J. M. Staniforth of the Western Mail.

The game saw the defeat of the new county cricket team, but their performance brought a smile to many of the faces of Cardiff's sporting community and those who had advocated the use of the Arms Park for recreational activities. Sadly, Frank Stacey was not among those in the Cardiff and County Club to congratulate or commiserate with the Club's member Edmund David, who had led the Glamorgan side into the field for their inaugural match. David had been a member of the Club since 1885 – the year when Stacey had died at his home at Llandough Castle following complications brought about by his ever-worsening asthma. How Stacey would have loved to have discussed this and other games with David, who went on to play on several more occasions for Glamorgan, including in the late 1890s following their admission into the Minor County Championship.

Educated at Cheltenham College, Edmund David was among a new wave of Club members who were very active in the town's sporting community. The son of the Rector of St. Fagans was also very busy in his work as land agent for the Margam Estate, but he found plenty of time to visit the Club, besides using the racquets and fives court in Westgate Street, or playing lawn tennis with family and friends on one of the courts laid out along the eastern boundary of the Park.

Edmund David (far right), with his parents.

Many other young gentlemen of the Club followed a similar pattern in their leisure habits when in town over the summer months. During the winter months they spent time in the Club's reading room studying form in the *Sporting Life*, checking the racing results which were posted on a notice board in the Club's lobby, or discussing other equine matters before heading off to the races, a point-to-point or a meeting of the local hunt. With business talk still frowned upon, chatter about all sorts of sporting topics filled the air in the Club. And there was plenty of excitement in January 1886 when the Welsh rugby team once again played an international at the Arms Park. This time they lost to Scotland, but the appearance of the men in red jerseys in the town, which was now home to 120,000 people, was further evidence of the continued rise of Cardiff.

But there were a few clouds hanging over the Cardiff and County Club, as well as their neighbours The Royal Hotel. For several years, there had been talk of the hotel expanding and, in March 1882, the Club's old premises in the hotel were put up for auction. No offers were made, so the lot was withdrawn, but it was clear to the Club's Management Committee that they

Cardiff and County Club.

DINING ROOM TARIFF.

BREAKFASTS.

Plain	1 - 6 -
Broiled Ham, Cold Meat or Eggs	2 - /-
Chops, Steaks, Fish, &c.	2 - 6 -

LUNCHEONS.

Soup and Bread	- 10 -
Bread or Biscuits and Cheese, and Glass of Ale	- 6 -
Sandwich and Glass of Ale	- 6 -
Plate of Cold Meat, Bread, and Glass of Ale	1 - - -
One Chop, Bread, Potatoes, and Glass of Ale	1 - /-
Second Chop	- /6 -
Steak, Bread, and Potatoes	1 - 6 -
Cold Meat, and Bread and Cheese	1 - 6 -
Vegetables	- 4 -
Salad	- 4 -
Glass of Sherry	- 4 -

DINNERS.

Joint or Steak and Vegetables, (including Ale)	2 - 6 -
Soup or Plain Fish, Entrée or Joint	3 - 6 -
With both Soup and Fish	4 - , -
Game extra.	

TEAS.

Cup of Tea or Coffee, with 2 slices of Bread and Butter or 2 Biscuits ...	- 6 -
Plain Tea	1 - 6 -
Do. with Eggs	2 - /-
Do. with Cold Meat, Chop, Steak, or Ham	2 - 6 -

needed to take action to secure the Club's long-term future. John Morgan and subsequently William Harvey each oversaw the discussions which resulted in the Club acquiring the freehold of the premises at 67 St. Mary Street for the sum of £1,500 in May 1883.

The Club therefore was in a decent position, should an offer be made to acquire their premises, but there was still plenty of talk around the town about the hotel once again putting up for sale some of their premises. There was much speculation about what might happen if a sale was agreed and new owners wanted to expand the premises and buy out the Club. It was clear to the Management Committee that the Club's future did not lie in the premises next to the Royal Hotel, and various alternative locations were considered. Their fears were confirmed when in March 1888, the hotel, plus its stabling block at the rear, was put up for sale at £45,000, and then two months later its parent company, The Cardiff Hotel Company Limited, went into voluntary liquidation.

Elsewhere in Cardiff, a host of new, grand buildings were being erected, reflecting the civic pride and identity of a number of flourishing companies and organisations. This was especially the case in the booming docks where in 1874 a Mercantile Club had been established in Rothesay Terrace, situated opposite the Chamber of Commerce and, according to the *Cardiff Times*, 'specifically for the patronage of gentlemen and merchants frequenting the area'.

It was a relatively small property and certainly not as grandiose as the Cardiff and County Club. But within a few years work had begun on the design of a Coal Exchange to act as the focal point for those engaged in the buying and selling of the 'black gold'. From 1886 they could use the lavish new building in Mount Stuart Square which was designed by Edwin Seward, the talented architect who had worked on the design of the Racquets and Fives Club in Westgate Street. Besides a large trading floor for the dealers, plus adjoining offices and rooms, Seward's plans included a clubroom where the traders could relax. For the first time, there was a rival to the Cardiff and County Club.

William H. Lewis, the newly-appointed Secretary of the Club, and the rest of the Management Committee clearly had plenty of matters on their minds and topics to mull over. Lewis, who had a solicitor's practice in Queen Street, had a more go-ahead attitude than Harvey, as typified by his dealings in securing a replacement for Charles Chalk. Given the expanding membership and the difficulties of the previous incumbent in undertaking all of the duties, he persuaded the committee to split the role between Edmund Parsons who acted as Manager, and Jacob Newton who was employed as a Senior Steward. The latter though did not last long, and returned rather disillusioned to Chelsea in London, having sent a letter to the committee stating 'my health entirely broke down through having to remain up until four, five or six o'clock in the morning, often on Sunday mornings, attending on members who chose to remain gambling at cards. This state of things being brought to the notice of the Chairman, Mr. George Williams, he directed me to turn the lights out at one o'clock every morning, which order I carried out, and of course, got disliked for doing my duty'.

Besides being an interesting commentary on Club life, it showed how busy and successful the Club had become. But with rival and opulent facilities soon to be available in a landmark building in the docks, the Management Committee considered whether they too should have a distinguished and much more spacious property of their own where the long-term future of the Club could be secured. St. Mary Street and the Royal Hotel had been a suitable location for the past 20 years, but with the developments along Westgate Street – whose recreational facilities were frequently used by Club members – another short move to a plot of land immediately adjoining the Arms Park where a new and larger building could be created seemed eminently sensible.

Indeed, in September 1887 the *Weekly Mail* and *Western Mail* newspapers each reported how 'a scheme has been promulgated for locating the Cardiff and County Club in new premises. For some time, the present building in St. Mary Street has been found to be too confined for the convenience of members. The exact locality of a new site has not yet been decided upon'.

Having made the decision to move, the Management Committee was delighted to hear from several businesses who were happy to acquire the Club's old premises and to develop them for commercial use, especially as the sale would provide the capital for a new building. With trade in Cardiff still on the up there was plenty of interest from a range of concerns. Eventually, in September 1890 an agreement was reached with the Wiltshire and Dorset Banking Company, and the three Trustees – Charles Williams of Roath Court, Colonel John Tyler of St. Hilary and Richard Wyndham Williams – oversaw the sale for the princely sum of £12,000.

As negotiations for the sale of the old premises were progressing smoothly during 1889, Lewis and his colleagues successfully sounded out the Marquess about a plot of land adjacent to the Racquets and Fives Club where, in the words of the *Weekly Mail* for April 30th, 1887, 'The pretty little ground used by the Quoits Club was sited'. The news of their acquisition came as something of a bolt out of the blue for the Quoits Club, but at least the nature of their rinks meant that they could be easily relocated elsewhere close to Westgate Street.

With a host of Bute employees on the membership roll of the Club, word quickly got out about the new location of the Club and there was soon plenty of chatter in the old premises about the move, the design of a new

Westgate Street, looking south, as seen in the mid-1880s with Jackson Hall in the foreground behind which is the land where the Clubhouse was erected.

A postcard from 1904, looking north, showing Westgate Street (left) and St. Mary Street (right).

building, and about the facilities the new premises might have. Discussions continued with the Bute Estate and, during the late summer of 1890, a verbal agreement was reached with the Marquess for a 99-year lease, and an annual rent at 50 guineas a year. This was formally signed off on September 29th, 1890 with the Club's Trustees, George Frederick Insole of Fairwater House, George Crofts Williams of Llanrumney Hall, and Colonels Trevor Bruce Tyler of Llantrythid and John Tyler of St. Hilary, acting as signatories to the agreement.

Following the verbal agreement, a series of possible designs were drawn up by Edwin Corbett, the chief architect of the Bute Estate, whose brother John had been a founder member of the Club. A few weeks later, a suitable design was agreed by the Bute Estate and the Club's Management Committee. The *Western Mail* initially described the proposed new premises as 'a handsome and substantial building,' and added a few months later: 'the new county club promises to be another important addition to the street architecture of Cardiff'.

Corbett's plan was for a three-storey stone building, with terracotta facings, mullioned windows and a covered entrance on to Westgate Street. The main feature of the ground floor was (and to many people still is) the magnificent dining room, the windows of which opened out on to a balcony overlooking the Arms Park and the rugby ground. Alongside was a large lounge and a waiter's room, from which members would be served drinks, and in the basement was the committee room and the Secretary's office, together with a scullery, two wine cellars, and a beer store. The

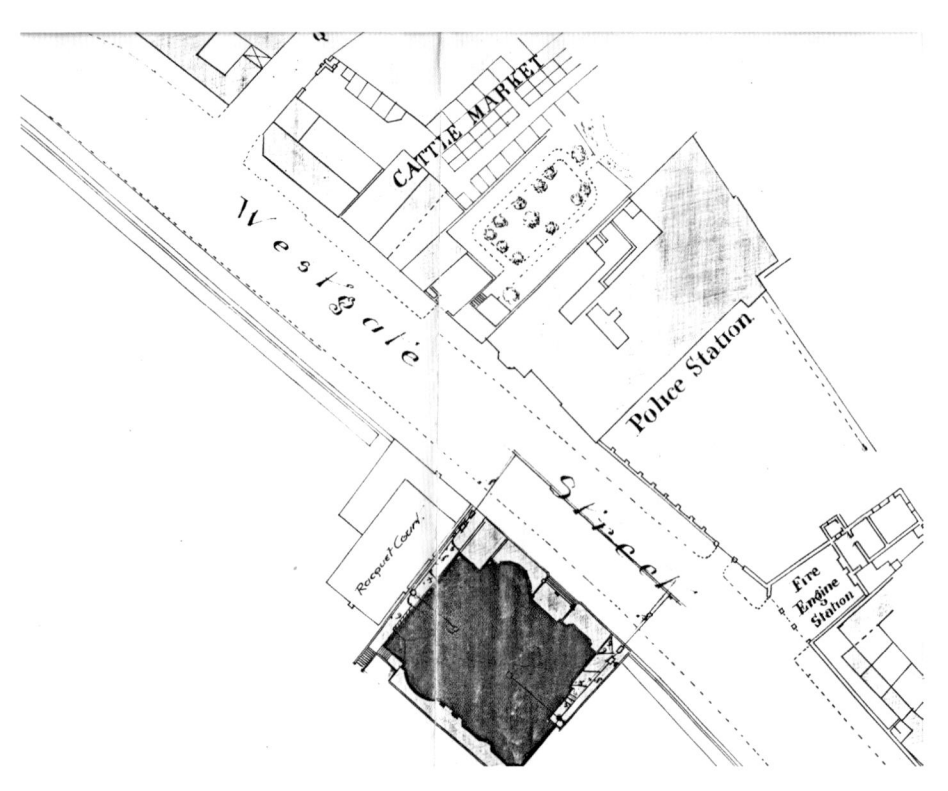

An extract from the contractor's plans from the 1890s, showing the site of the new Clubhouse and its neighbours on Westgate Street.

upper floors included further spacious lounges, with a card room, billiards room and a reading room, and in the attic was space for six bedrooms and a living room for the use of the servants.

These plans echoed, very much, the character and ethos of the Club, in its combination of respectability and flamboyance. Its grand scale and Renaissance style also symbolised stability and power, and the use of red sandstone dressings and the installation of verandas reflected boldness and imagination. The plan received the green light from the Bute Estate in August 1890, before planning approval was secured from Cardiff Council on October 13th, 1890.

Tenders for the construction work were also issued and the eventual appointment was Samuel Shepton, a builder based in Dumfries Place whose portfolio of works included the *Western Mail* buildings at the southern end of St. Mary Street, the Catholic Grammar School in David Street and a series of properties in The Royal Arcade. His initial progress however at the Westgate Street site was slowed during 1891 since he had to secure £3,000

NEW COUNTY CLUB FOR CARDIFF.

For some years the members of the Cardiff County Club have had their headquarters at the northern end of the Royal Hotel, St. Mary-street. When, however, the Royal Hotel Company decided to enlarge the hotel, the members of the club were asked to look out for new premises. This was done, and, after considerable negotiations, Lord Bute granted a site in the Cardiff Arms Park, adjoining the tennis club premises, and on the ground occupied at present by the Cardiff Quoit Club. Mr. E. W. M. Corbett, architect, was entrusted with the preparation of the plans, which provide for the erection of a handsome and substantial block of buildings. The new club, the elevation of which is shown in the above sketch, will be three storeys high, and built with local stone with terra-cotta facings. It will be lighted by heavy mullioned windows, and the entrance, which is in the centre of the structure, will be approached by a flight of steps. The door opens into a vestibule, with porters' offices on the right. On the left will be placed the committee and secretary's room, 20ft. long by 13ft. in width. Passing through the hall, on the right will be a spacious reading-room, 27ft. by 21ft., and immediately opposite, in the back part of the premises, a similar room. At the extreme end of the hall will be a magnificent dining-room, 45ft. by 23ft., the windows of which open into a balcony overlooking the park. The balcony will run the entire length of the building. On the same floor will be the manager's office, still-room, serving hall, and the usual conveniences. In the basement of the building splendid kitchen and scullery accommodation has been made. At the top of a noble flight of stairs, which communicates with the first floor, will be a large card-room, immediately over the front reading-room. Over the dining and back reading rooms will be two handsome billiard-rooms, lighted by the windows overlooking the park and from two large skylights in the flat above. Over the committee-room will be various dressing-rooms and other offices. The attic plan shows provision for six large bedrooms. It is expected that the work of erection will be proceeded with in a short time.

How the Western Mail reported the news about the proposed new Clubhouse on August 30th, 1890.

An image of the Clubhouse from The Builder, March 13th, 1897.

of additional expenditure for pile-driving and laying additional foundations in what had been the old bed of the River Taff.

During the late spring and early summer of 1892, his progress was disrupted again by an industrial dispute which affected all building contractors across Cardiff. Fortunately, some of the plasterers and other decorators involved in the scheme were directly employed by the Butes so they were able to continue working, but the labour unrest and work to rule meant that it wasn't until the late summer of 1892 that the Club members were finally able to use their new headquarters. The *Evening Express* of August 12th, reflected the mood of great delight by reporting how the Club 'was just completing the move to the larger and more commodious building which has now been erected'. A new chapter in the history of the Club was about to unfold.

7

Berkeley and the Brains

"We are a Club with a tradition of sociability, mutual respect and good manners."

When the new premises opened its doors for the first time, Richard Wyndham Williams had been President ever since the Club's inauguration in 1866. During his lengthy and distinguished term in office, he had twice seen the Club outgrow its home and move to new buildings. As he welcomed members into their new base in Westgate Street during the late summer of 1892, and presided over a grand opening function, the venerable solicitor, now in his 65th year, hoped that the Club would enjoy many years of stability and permanency.

This duly came about, largely because of its strong relationship with the Bute Estate, as well as the actions of Charles Berkeley, the Club's Secretary from 1891. Though there was no direct stipulation that a sympathiser of the Marquess should be at the helm of the Club, it certainly helped that Berkeley was another Bute-man, being the nephew of John Corbett, one of the key figures in the team of land agents, solicitors and administrators working for the Marquess. Berkeley, an Essex-born accountant, oversaw the final round of arrangements for the new building, as well as the negotiations for the lease, with his friends and colleagues in the Bute Estate.

Charles Montague Berkeley was the son of the vicar of Southminster and, after attending Mount Radford House School in Exeter, he trained to be an accountant and with his wife Emily, moved to Cardiff in 1876 to work under his uncle in the

Richard Wyndham Williams – the inaugural President.

Bute offices. Charles also joined Cardiff CC, and within a year or so was playing in the 1st XI with great effect, taking six wickets with his spin bowling in 1878 in the victory over the Sneyd Park side, one of the crack teams in the Bristol area. In August 1881 he was given the task by the Bute Estate of organising the sporting activities which took place in Cathays Park to celebrate the birth of the Fourth Marquess.

He carried off these duties well but, shortly afterwards, he moved away to work in Rutland, before returning to Cardiff in 1884 and setting up his own business as an independent accountant and, just for good measure, joining the Club. He also secured a decent home at The Wynyards in Plymouth Road, Penarth. In the next few years he performed several roles including, at various times, Secretary of the Park Hall Company, Secretary of the Cardiff Proprietary School Company (based in Dumfries Place), Secretary of the Roath Sanitary Steam Laundry (in Marlborough Road), and President of the South Wales and Monmouthshire Society of Incorporated Accountants and Auditors.

Charles was also a noted breeder of dachshunds and, besides being Chair of the Cardiff and South Wales Kennel Club, for over 20 years he was Secretary and general factotum of the Cardiff and South Wales Horse

A group of members standing on the steps leading to the garden, circa 1897.

Charles Berkeley.

Show, the largest provincial show in the UK, held each year on the Bute-owned recreation ground in Sophia Gardens. Having such a well-known figure in the sporting and social world of south Wales was of great benefit to the Club, and he methodically carried out his duties as Secretary until the 1920s. By this time, the new premises in Westgate Street were a roaring success and to reward his strenuous efforts in ensuring the efficient and productive move from St. Mary Street, Charles, or 'Uncle', as he was known to everyone in the Club and the sporting world of south Wales, was awarded Honorary Life Membership. He continued to be a familiar face around the Club for many years and, given his love of numbers and cricket, it was natural that he should take on the duties of scorer for Glamorgan CCC, and he was still in post in May 1932 when he died at his home in Whitchurch.

Sir Edward Stock Hill.

It was testament to his efforts, and the support of the Bute Estate, that the Club, like the town itself, went from strength to strength in the early Edwardian era. Several of the shipping magnates and other notables from the docks regularly visited the Club in Westgate Street, and thoroughly enjoyed the ambience, as well as the superb facilities of the new premises. Among the leading figures from the life of the docks were Sir Edward Stock Hill of Rookwood House, and his sons. Sir Edward was the MP for Bristol South between 1886 and 1900, and his family oversaw the effective operation of a dry dock in Cardiff, and a railway, the West of England Line.

Another prominent Cardiff ship owner and member of the Club was Henry Clay of Piercefield Park in Chepstow. He had moved from Derbyshire having made his

Jack Brain, seen here during his period as a young cricketer with Gloucestershire.

money in banking and brewing, before diversifying his interests into shipping. His son, Charles Leigh Clay, born in 1867, oversaw these operations for his father and was a member of the Club from 1890. Their line, Claymore Shipping, subsequently became one of the successful ventures in the inter-war era. By this time, William Reardon-Smith, another Devonian, had also established a lucrative shipping business from the docks and was active in life at the Club. Established in 1905, his shipping interests went from strength to strength either side of the Great War and, by the time of his death in 1935, he owned 28 vessels. As befitted one of Britain's leading ship owners, he was a generous philanthropist and benefactor, especially to the National Museum of Wales to which he donated around £150,000 between 1915 and 1935.

During the 1880s a rival to the Bute-owned docks emerged at Barry. Promoted by David Davies, the coal magnate of Llandinam, and served by newly-built railway lines from the Rhondda, they had the advantage over Cardiff as they could be used even at low-tide, and charged lower fees than those levied by the Bute Docks Company. By 1901 exports from Barry exceeded those of Cardiff, and David Davies boldly claimed that 'grass would grow in the streets of Cardiff' as trade shifted to Barry. But Cardiff maintained its pre-eminence with the Coal Exchange in Mount Stuart Square remaining the place where the price of coal on the British

Jack Brain walks off the Arms Park cricket pitch after being dismissed in a game for Glamorgan in 1903. The Clubhouse and Jackson Hall are visible in the background.

Fred Insole

Fred Insole was the Club's President during the years leading up to the Great War. Born in Cardiff in 1847, he lived at Ely Court and owned several collieries in the Taff Valley. In addition to overseeing the export of coal from his mines via the docks in Cardiff and Barry, he also served as a Director of the Taff Vale Railway.

Fred was the archetypal country gentleman enjoying a range of field sports including hunting and shooting, and he also loved playing cricket. Indeed, his family had been instrumental in the creation of Fairwater Cricket Club which played in the grounds of the family's house adjacent to Fairwater Road as well as on land which now forms the site of Cantonian High School.

His extensive social connections, rather than any outstanding abilities at cricket, led Fred to being chosen for the Glamorgan team in 1874 and 1875. What he lacked in talent was more than matched by great enthusiasm and he duly oversaw a number of very successful cricket matches for his friends from the Club at the Fairwater Cricket Club.

During the winter months he also organised a number of hunting parties for his sporting friends, who included both Jack and Sam Brain, as well as assisting with the arrangements for point-to-point steeplechases. One of his closest friends was Edward Stock Hill (later Sir Edward) who lived at Rookwood House. The Bristol-born gentleman was the son of the founder of the Hill's Dry Docks in Cardiff, besides being a Founder Member of the Club in 1866.

Fred Insole.

Sir Edward shared Fred's passion for cricket as well as hunting and, despite his duties as Conservative MP for Bristol South between 1886 and 1900, he was a regular diner in the Club, taking every opportunity to meet up with Fred, the Brain brothers, and his other pals in order to catch up on the latest gossip!

market was determined, as well as being the venue for the first-ever million pound deal in 1907.

Cardiff's position as the economic epicentre of the coastal plain was enhanced by the diversification in its industrial portfolio, following the decision of the owners of the Dowlais Ironworks in Merthyr, who later became part of the Guest, Keen and Nettlefolds enterprise, or GKN, to construct a steelworks close to the docks at East Moors, thereby saving money and time on transporting products to and from the valleys.

The steelworks was formally opened by the Marquess of Bute on February 4th, 1891 and its creation saw, over time, a shift in the centre of steelmaking from the valleys down to the coastal fringe, with the trend being reinforced

The original pavilion at the Arms Park, seen during the match between South Wales and Australia in 1902.

by the opening further west of steelworks at Margam (Port Talbot), as well as subsequently at Llanwern to the east of Newport. The steel magnates, and the well-to-do agents who bought iron ore and sold steel were among a new wave of businessmen who became members of the Cardiff and County Club during the late 1890s and into the 1900s.

They were joined by another wave of English-born migrants who made a decisive influence on both the affairs of the Club, as well as on life in general in Cardiff in the Edwardian era. Amongst these were the Brain brothers, Jack and Sam, who had been sent by their uncle to oversee the running of the family's Old Brewery in St. Mary Street from the late 1880s onwards. Jack was a leading cricketer with Gloucestershire and having enjoyed a decent career at Clifton College and subsequently Oxford University, many thought he would succeed the legendary Dr. W.G. Grace as captain of the West Country side.

But his uncle, Samuel Arthur, had acquired The Old Brewery in 1882 and, in the course of the next few years, installed his nephews in managerial positions. Jack moved to Cardiff during the late 1880s to learn the ropes and to start his career in brewing. He soon became active with the town's cricket club, besides taking over the captaincy of Glamorgan in 1891 and becoming a member of the Club in 1892.

Jack remained captain of Glamorgan until 1908, during which time it went from strength to strength and, by the first decade of the 20th century, had become one of the strongest Minor County teams – a transformation met with great acclaim by Jack's many friends in the Club, and entirely down to his skill and man-management. His business acumen also saw the Old Brewery undergo a series of successful transformations and, following the purchase of many taverns in the town, the family's brewery rivalled that of Hancock's as the leading producer of beer in south Wales.

In 1900 Glamorgan were joint winners of the Minor County title. With decent crowds at the home games at the Arms Park, and the quaffing of Brains ales in the town, Jack and his brother realised that what was good for Glamorgan was good news as well for Cardiff and south Wales. There was a great mood of optimism permeating the lounges and dining room at the Cardiff and County Club, where Jack and Sam relaxed with their friends from the sporting, business and political world, many of whom believed that the time had come for Cardiff to be given city status.

During the early 1900s the Brains, along with a number of prominent members of the Club, did their bit to enhance Cardiff's sporting reputation. A showcase cricket match was arranged, using Jack's influence within the MCC, at the Arms Park in 1902 as Glamorgan joined forces with Wiltshire to play the Australians. Some 12,000 people attended the game, with trains

Jack Brain and the wizard wheeze!

Potentially, it was the story on which Jack Brain could have dined out for the rest of his life. But the wizard wheeze of the Oxford University cricket XI of 1886 to scupper W.G. Grace's plans of filling his boots against the undergraduate attack by plying the good doctor with copious amounts of alcohol rather backfired and, instead, it became a good yarn which Jack and his brother Sam would regale, with typically self-effacing humour, to their friends in the Cardiff and County Club for many years.

It stemmed from the appearance of the great English batsman – and an iconic figure in Victorian Britain – for the MCC against the Dark Blues as the Gloucestershire batsman sought further landmarks in an already glittering career. "He's going to put us to the sword," said one of Jack's colleagues as the leading figures in the Oxford side met during the week before the game to contemplate their tactics and how to counter what was likely to be Grace's massive influence on the game. After much debate, they duly hatched a plan to try and get W.G. drunk on the first night of the game, with the hope that the effect of too much alcohol would reduce The Champion to a mere mortal.

Jack, who had been in the Gloucestershire side for the previous three years, duly agreed to host a dinner party for his county captain and other leading figures in the MCC party. "He'll never suspect your intentions," said one of Jack's friends with a glint of mischief in his eye. After W.G. had posted an unbeaten 50 before the close of play on day one of the match, the students politely applauded Grace as he left the field, before quickly getting changed and making their way to Jack's party.

W.G. jovially consumed glass after glass of champagne but, in the words of a contemporary, "he had a massive constitution and a bottle or even two bottles of champagne had no effect on his mighty frame." Towards the end of the evening, W.G. sensed that he was being set up and feigned that he was becoming increasingly merry. The following morning he deliberately missed the ball several times in the nets, and the college closes and thoroughfares of Oxford were soon abuzz with rumours that "W.G. was tight last night and he still can't see the ball this morning".

When play resumed at The Parks there was an above average gathering, with many interested onlookers hoping to see if the prank that Jack and his chums had engineered would bear fruit. But right from the opening overs, it was clear that the little pantomime was over, as

> W.G. effortlessly eased towards his century before he proceeded to make inroads into the Oxford batting. Indeed, it was some of the undergraduates who appeared to be far worse for wear after their excesses the previous night at Jack's party, and those who thought they would see England's greatest cricketer play in an inebriated state saw him take all ten Oxford wickets for just 49 runs, as the undergraduates were bundled out for 90.

packed with excited cricket supporters from as far away as Milford Haven in the west and Swindon in the east. Many pints of Brains beer were supped in the taverns in the town and, in the Club, some members were able to watch events from the balcony at the rear of their building.

The game coincided with the coming-of-age celebrations of John Crichton Stuart, the Fourth Marquess of Bute. As well as a special dinner in the Club, a number of other events were held when the Australian cricketers were in town, including a garden party in the Castle Grounds where the famous Australian cricketers, together with leading members and officials

A postcard, dated 1904, showing the new buildings at the southern end of Westgate Street.

of the Glamorgan club, were all able to mix with the great and good of south Wales society and Tory sympathisers.

Although the game was won comfortably by the tourists, it didn't dampen the enthusiasm of the Brains or the Glamorgan committee, as they oversaw the creation of a grandiose pavilion at the Arms Park, a building that reflected the ambitions and lofty dreams of Glamorgan CCC. In the autumn of 1904, they approached the MCC about allocating the first Test Match in the 1905 series with Australia to Cardiff. By the slender margin of a single vote, Nottinghamshire won the day, but the MCC officials had been so impressed by Glamorgan's claims that they allocated a fixture to Cardiff during the August Bank Holiday of 1905, with the Australians playing the cream of talent from south Wales.

Once again, a bumper crowd turned up at the Arms Park, with the public houses, like the bar at the Club, doing a roaring trade, but the match resulted in another win for the tourists. However, all concerned were mightily impressed and it proved be the first of a series of momentous events in Cardiff during 1905.

8
Chukkas and Putters

"A meeting place for like-minded people to meet, talk, read, drink and eat."

By the turn of the century, there were sufficient members in the Cardiff and County Club with an interest in all things equine for a polo club to be formally created. It followed a series of informal gatherings and, with Charles Berkeley's encouragement, plus his wealth of contacts, a formal club known as The Cardiff and County Polo Club, was formed in 1904. Its leading figures were Captain Lionel Lindsay, the Chief Constable of Glamorgan, and Walter Shirley, an eminent solicitor. Lindsay was the second son of Lieutenant-Colonel Henry Gore Lindsay and the Hon. Ellen Sarah Morgan, the daughter of Lord Tredegar. Born in Brecon in 1861, Lionel Lindsay grew up at the family's home, Woodlands, in Leckwith before a brief period in military service. He then joined the Glamorgan County Police where he rose to the rank of Chief Constable.

Lindsay had joined the Club in 1891 and, during the next few years, one of his closest friends and partners on the polo field was Walter Shirley of Plasnewydd House in Roath. The son of Lewis Vincent Shirley, another well-known member of the Club, he had been educated at Malvern, before reading law at Trinity College, Oxford. He then joined his father's practice with William Luard in Castle Street, which was closely affiliated with the Bute Estate. Walter grew up at Plasnewydd House, a large property dating from the 1780s which lay to the east of what is nowadays City Road. Prior to 1905, it was known as Castle Road, largely because of the castellated features of the mansion known as Roath Castle, with the Plasnewydd Estate itself belonging to John Matthew Richards, a prominent alderman of Cardiff and a key figure in the Bute Estate.

The Plasnewydd Estate had extensive grounds, as well as Ty'n-y-Coed Farm to the north, where cricket and lawn tennis also took place. Polo may well have also been played at Plasnewydd, as Lindsay and Shirley were both keen riders and huntsmen, and participated in many of the equine activities organised by the Williams family of Roath Court. They also rode regularly with the Glamorganshire Hunt, besides competing in local point-to-points. Their friendship was such that Shirley later acquired the Lindsay's family home in Leckwith and it was at Woodlands where Shirley lived for many years with his wife Eliza, the daughter of Archibald Hood, a prominent colliery proprietor in the Rhondda and supporter of the Barry Dock and Railway scheme who was another, very well-known Club member.

Polo had increased in popularity around the turn of the century, both as a form of manly exercise and a suitable form of socialising for gentlemen, so Lindsay and Shirley discussed with their close friends in the Club the feasibility of forming a polo club. They gained the support of other leading members of local society, including Lieutenant-Colonel Evelyn Orlebar, a Royal Artillery officer born in Bedfordshire who had moved to Cardiff following a posting to Maindy Barracks.

Another key figure was Other Robert Windsor-Clive, the eldest son of the Earl of Plymouth who lived at St. Fagans Castle. Born in Fordbigge in Worcestershire in October 1884, Windsor-Clive was another enthusiastic huntsman and energetic polo player but, tragically, his association both with the Polo Club and the Cardiff and County Club was all-to-brief as on December 23rd, 1908, he died of enteric fever in Agra when serving as *aide-de-camp* to the Viceroy of India. News of his passing reached St. Fagans on Christmas Day, and the sadness of the villagers was matched by that of Club members when they read the sad details in the Boxing Day newspapers.

Windsor-Clive had proved to be a very useful ally for the polo-loving members of the Club because he helped persuade his father to lease to the Polo Club an area of 14 acres at the rear of Island Cottage to the west of Merthyr Road in Whitchurch. Indeed, the Earl's willingness to make land available may well have been the catalyst for the formalisation of arrangements for polo matches and practice, and the creation a proper club.

The field offered by the Earl was just a ten-minute walk from Llandaff Railway Station and it was here, from 1904 onwards, that the Club's members took part in a series of matches against some of the leading polo teams from south-west England, as well as teams from various hunts. The Polo Club also organised gymkhanas and pony races on their land in Whitchurch, allowing many ladies and girls to participate, besides socialising in a convivial atmosphere with the young gentlemen. The earliest activity took place in May and June 1904 as a series of trial games took place on the Whitchurch Polo Field, ahead of the Club's inaugural fixture against the North Wiltshire club. During the final week of August, a five-day tournament was held with invitations to take part being accepted by the Blackmore Vale Hunt Club, the North Devon Hunt, plus polo teams from Stratford-on-Avon and Chippenham.

The success of these inaugural ventures led in April 1905 to a pavilion being built at the Whitchurch Polo Field, with the wooden structure comprising two changing rooms, where the gentlemen could safely leave their clothes, as well as a refreshment room. Another invitational tournament took place in August 1905 and, the following year, members of the Cheltenham Polo Club met Cardiff and County Club members at the Whitchurch Field.

Through Charles Berkeley's influence, a series of polo fixtures for Club members were added in 1910 to the programme of events at Cardiff Horse Show, which was held on the Recreation Ground at Sophia Gardens. By this time, one of the most active members of the Club's polo team was Sam Brain, the younger brother of Jack, who had also moved to work in the family's brewery after completing his education at Oxford. A Club member since 1894, Sam was also a county cricketer with both Gloucestershire and Glamorgan, besides enjoying hunting and taking part in shooting parties. He also rode in various point-to-points across the area, and in the Club's annual steeplechase, as well as riding the horses owned by himself and his brother.

A curious hat-trick

Only one man so far in the history of first-class cricket has completed a hat-trick of stumpings – Sam Brain. He achieved this feat when keeping wicket for Gloucestershire against Somerset in 1893, shortly after coming down from Oxford, where his deft glovework had impressed the county's officials who were looking to inject new blood into what had become a rather moribund side.

His feat came in the opening game of the 1893 Cheltenham Festival, in a match which also featured in the Somerset line-up Vernon Hill, the son of Edward Stock Hill of Rookwood House, Conservative MP for Bristol South and another well-known figure in the Cardiff and County Club. The Gloucestershire side also included 16-year-old Charles Townsend, another cricketing prodigy from Clifton College, and it was the young leg-spinner – in only his second county match – who, together with Brain, wrote a name for themselves in the cricket record books shortly before the end of the second day's play.

By this time, the visitors had amassed a lead of 301 and, with a declaration likely the following morning, there was something of a carefree mood in the Somerset camp as Townsend began his eighth over ten minutes before the scheduled close of play. The youngster probably thought he might get another over, but as it turned out it was the last over of the day. To the fourth delivery of the over, Arthur Newton advanced down the wicket, played an ungainly heave, missed the ball and was promptly stumped by Brain. To the next two balls, the Somerset professionals, George Nichols and Ted Tyler, also went down the wicket to the young bowler and departed in identical fashion 'stumped Brain, bowled Townsend.'

There was much celebration out in the middle as Sam was warmly applauded for his efforts as he walked towards the pavilion. There were broad smiles as well on the faces of Sam's friends in the Somerset side and, together with Hill, he headed off into Cheltenham town to toast his record-breaking achievements. Other members of the Gloucestershire side followed the pair and – quite understandably – they all had a little bit of a headache the following day as they were dismissed for 174 runs to give Somerset victory by 127 runs.

Sam Brain.

The connection between the Brain family and both Glamorgan CCC, as well as the Cardiff and County Club was maintained by his sons Pat and Michael who each kept wicket for the Welsh county and were regular faces in the clubhouse either side of the Second World War. After attending Cheltenham College, Michael's son Chris continued in the family tradition by joining the family brewery and rose to the position of Chairman before retiring in 2009. Chris maintained another family tradition of keeping wicket and, during the 1980s, he played in a South Wales Hunts' match against the Army during which he kept wicket to the bowling of Jeremy Townsend, the grandson of Charles.

Another enthusiastic member of the Club's polo team was the Rev. Hugh Jenner, the grandson of founder member Robert Jenner of Wenvoe Castle. Hugh Jenner was a decent polo player and, after completing his theological training at Oriel College, Oxford, he enthusiastically turned out for the Cardiff and County Club's team. Born in July 1872 in Herefordshire, Hugh also formed his own team, and maintained his links with the Club, and south Wales as a whole, after taking curacies at Falfield in Gloucestershire and near Taunton in Somerset.

He also took a keen interest in golf and, after taking over the management of the family's estate, he provided land for Whitchurch Golf Club in 1914, as well as for the Wenvoe Castle Club which was laid out in the grounds of his family's home in 1936. His support of the Whitchurch gentlemen followed an approach by a consortium led by Edmund Howell, a Pembrokeshire-born chemist who lived in Colum Road and, like other members of the city's growing middle-class, was looking for a site where respectable recreation could take place.

Herbert Thomas, the influential owner of Melingriffith Tinplate Works, and another member of the Club, had already given the group a £100 loan to allow negotiations to commence with various landowners. Howell and his friends also had the support of another sporting clergyman, the Rev. Thomas Ewbank, a Wesleyan Methodist minister who had read theology at Christ's College, Cambridge, before holding curacies in Lincolnshire, Lancashire, Sussex, Huntingdonshire and Nottinghamshire. Born in Richmond, Surrey, in 1847 he had been a talented cricketer in his youth before taking up golf in later life.

Ewbank agreed to act as Chairman of the Whitchurch Golf Club and, from his home in Velindre Road, he helped oversee the negotiations for a course that was easily accessible by public transport. The group's attention was focused at first on plots of land at Mynachdy and Gabalfa near to the tram terminus at Maindy Barracks in what is now North Road. Sites at the Wenallt and the Graig were also considered but their remoteness was seen as a potential stumbling block. But a far more suitable alternative arose after Hugh Jenner agreed in 1914 to lease to the Club land to the east of what is now Manor Way. At the time, the area was part of Pentwyn Farm and had initially been earmarked for housing development as Jenner gave his backing to the group of academics and supporters of the Garden City Movement who were trying to emulate the achievements of Ebenezer Howard at Welwyn and Letchworth by creating a similar leafy suburb at Rhiwbina.

By Christmas 1913 only 34 properties close to Rhiwbina Halt had been completed by the Cardiff Workers Co-operative Garden Village Society. With

An extract of the Club's Licence Book from 1913 to 1921, highlighting the impact of the First World War on membership numbers.

doubts over the sources of financial support to allow hundreds more to be built, Jenner decided, perhaps after some persuasive words from Thomas Ewbank, to lease the land to Whitchurch Golf Club and the initial nine-hole course was formally opened in mid-May 1915. Over a 100 years later, the connection between Whitchurch Golf Club and the Cardiff and County Club remain strong with, in 2016, Honorary Secretary Ceri Preece serving as golf club captain.

Rhiwbina was not the only place in the Cardiff suburbs where supporters of the Garden City Movement were active during the late 19th century, as John Cory – one of the most successful entrepreneurs at Cardiff Docks – was working to the west of the city to create such a settlement near his substantial home at Dyffryn House. Known variously as Coryville, Coryndon and Glyn Cory, the scheme envisaged an area of mixed housing and a golf course on land half a mile from Peterston-super-Ely railway station.

His father, Richard Cory, had made his fortune after moving from Bideford in North Devon and establishing, in 1831, a ship chandler's business in Cardiff. A ship brokerage soon followed, and a coal-exporting business as the Bute Docks opened. Together with his brother Richard junior, John diversified the company's interests, besides opening coal depots overseas in countries including Egypt, India, China and South Africa. From 1868 the Corys also owned collieries in the Rhondda Valley and, during the closing years of the 19th century, they supported the group overseeing the creation of Barry Docks.

Richard Cory senior was a staunch Methodist and teetotaller so, whereas others based at the thriving docks enjoyed socialising at either the Exchange or the Club, he largely stayed away. John was encouraged by his Liberal friends to become a member of the Club, but his fervent support of the Barry cause appeared to ruffle more than a few feathers, and in 1882 the local newspapers carried articles, purporting to voice the concerns of his supporters over the way he had been blackballed when applying to join the Club.

Relations with the Cory family became less strained with John's son Clifford being admitted to the Club in 1890. Clifford followed his father into the coal trade and, until his death in 1941, he played a prominent role in the industrial and political world of south Wales, serving as Chairman of Cory Brothers, besides being President of the Cardiff Incorporated Chamber of Commerce in 1907 and 1908. From his home at Llantarnam Abbey he served as High Sheriff of Monmouthshire in 1905, and represented Ystrad on the Glamorgan County Council. Indeed, he became very active in Liberal politics and went on to become MP for St. Ives in addition to being a prominent member of David Lloyd George's Coalition government between 1918 and 1924.

In his youth, Clifford Cory had also been a fine polo player. Besides being an active member of the Club's polo team, he was a member of the Hurlingham Club, and in 1909 he was chosen for the England polo team which met their French counterparts at Cannes. The year before, Cory had also organised a match for the Club team against his own side from Llantarnam. Judging by the reports in the local newspapers, it was one of the highlights of the social calendar in the summer of 1908, with the contest being watched by Lady Bute and other notables from the Castle Estate. Given the Butes' support of the Tory cause in Cardiff, the polo match was also – like so many things in the town during the Victorian and Edwardian era – a lively contest between Conservatives and Liberals!

Members enjoying a Club outing to Bristol in the 1920s.

9
1905 and all that

"A place where members always feel at home."

On Sunday, October 22nd, 1905 the Mayor of Cardiff, Alderman Robert Hughes, received a letter from Whitehall which confirmed that King Edward VII had formally conferred city status on Cardiff. The news was greeted with an outpouring of civic pride and much ceremony by the corporation's officials, and the many members of the Cardiff and County Club who also held positions within the council and other governing bodies. Great celebrations took place outside the Town Hall, with the Mayor and the Marquess of Bute receiving a massive ovation from the huge crowd which had gathered in St. Mary Street. After some impromptu speeches, there was the singing of the Welsh national anthem and the sound of this, and other melodies, filtered into the Club and other nearby buildings.

Construction of a new town hall and law courts was already well underway at Cathays Park. In 1897 the tender had been awarded to the London-based firm of Lanchester, Stewart and Rickards, and some upgrading – to reflect Cardiff's new status as a city – took place ahead of the formal opening in October 1906. A number of lavish functions took place during the autumn of 1905 in the Club in Westgate Street, as the members of Cardiff's urban bourgeoisie celebrated their achievements.

All of the celebrations were overseen with the minimum of fuss by William Herbison, one of the longest-serving secretaries of the Cardiff and County Club. Born in Cork in Ireland in 1853, he and his wife Mary had plenty of experience of running clubs, having overseen the running of the Toxteth Club in Liverpool before moving to south Wales in 1898 and succeeding Charles Berkeley. The Irishman also lived at the Club's premises in Westgate Street, in the spacious flat and rooms on the upper floor. According to the 1901 Census, the other staff resident at the Club were Alfred Lane – waiter, James Donald – pantryman, Samuel O'Leary – page,

Polly Palmer – kitchen maid, Eleanor and Elizabeth Candy – housemaids, and Catherine Buxton – barmaid.

Other staff worked in the Club, including cooks and kitchen porters, especially on busy days such as those when rugby internationals were staged at the Arms Park. These remain among the busiest experienced by the Club and their staff; back in the 1900s the celebrations associated with some of the victories by the Welsh rugby team were momentous not just for the members of the Club, but for Wales as a nation.

Indeed, the late Victorian era had seen a steady rise in the number of rugby-playing members including Hugh Ingledew, a Cardiff-born solicitor, who specialised in railway law, and was very prominent in rugby circles in Edwardian Cardiff. His father John Pybus Ingledew had run a successful solicitor's practice for many years in the town, acting as the legal representative for the Taff Vale Railway. The Ingledews lived in Charles

A sketch of Sophia Gardens, drawn circa 1890, showing the place where many Club members would relax and promenade with their families.

Henry Shewbrook (sitting, front row, left) with his pupils and fellow staff at the Monkton House School. This photograph was taken during the 1880s and shows many children of Club members. In future years many of those photographed would become Club members themselves.

Street and subsequently Windsor Place before securing a larger and grander home in Cathedral Road. This was (and is) next to Sophia Gardens, and the area of tree-lined walks and ornamental gardens, opened by the Butes in 1859 as the first public park in Wales. To the north of the gardens was the large recreation field where Hugh and his many sports-mad friends played rugby in the winter and cricket in the summer.

The Sophia Gardens Recreation Ground was also used by the pupils of Monkton House School for recreation, with John Ingledew being among several parents who assisted the headmaster, Henry Shewbrook, in the coaching and administration of cricket and rugby. A new sporting community therefore evolved during the closing years of the 19th century with the young gentlemen like Hugh Ingledew, duly entering the professional classes, working in Cardiff, playing rugby and cricket for the town's club, and becoming members of the Cardiff and County Club where their sporting friendships and old school ties were both nurtured and preserved.

The Club, in the early 20th century, had several other former pupils from Monkton House, as well as old boys from the Reverend Green's Classical and Commercial School – a well-established rival, based in Charles Street – plus former scholars from Cardiff College, another private and fee-paying school. All of these educational institutions catered for the academic needs of the swelling numbers in the professional classes as well as other members of the urban elite, with Cardiff – according to the 1901 Census – now having 164,333 residents.

Hugh Ingledew read law at St. Edward's College, Oxford before returning to live and work in Cardiff, where he joined his father's practice. He qualified as a solicitor in October 1890 and worked with his father before he retired in 1893, with Hugh taking over all of the work on railway law, while his brothers Arthur and Norman, who had also joined the Cardiff and County Club, looked after the mercantile and commercial business of the practice. Despite his busy legal duties, Hugh found plenty of time for recreation. He played cricket for both the Cardiff club and the Garth team in Radyr, as well as in 1891 playing five times for Glamorgan where his athletic fielding and swift running more than compensated for the presence of a few rather more portly amateurs.

However, for Hugh Ingledew, like several of his Corinthian friends, it was at rugby that he really excelled. He had made his debut for the Cardiff 1st XV in the season of 1887-88 and after several seasons of success as a half-back in the Cardiff side, he was chosen in the Welsh side for the away match against Ireland on St. David's Day, 1890. The side was captained by Arthur Gould of Newport RFC and one of the early superstars of Welsh rugby. The fiercely contested game ended all-square but the match is more notable for the fact that a mass brawl broke out between the two teams during the after-match dinner in Dublin, and nine of the Welsh players ended up in court the following day.

Later in 1890, Ingledew became a founder member of the touring invitational rugby team, the Barbarians. It had been the dream of William Carpmael, a law graduate from Cambridge University, that playing rugby could promote good fellowship, through rugby tours, as well as encouraging good behaviour both on and off the field. In 1889 Carpmael had organised a tour with Clapham Rovers, in which the team faced five or six clubs in the Midlands and Yorkshire areas, before the following year organising a tour for a side called the Southern Nomads during which the Barbarians RFC were formed. From his father's business in Chancery Lane, in London,

Ralph Sweet-Escott.

Carpmael acted as secretary of the new rugby club, with Hugh Ingledew being invited to tour Northern England in December 1890 where matches were played against Hartlepool Rovers and Bradford. For Ingledew, it not only represented further recognition within the rugby-playing world but the links with Carpmael and other like-minded members of the legal world in London considerably assisted his family's practice in Cardiff, and led to Ingledew dining in the Club with notable lawyers when they were in town on business.

In the years leading up to the Great War, Hugh Ingledew became a member of the Council of University College, Cardiff, and in 1913 he came to prominence again when he was instrumental in establishing the School of Mines at Trefforest to the south of Pontypridd. As treasurer and secretary of the new institution, owned and funded by the major Welsh coal owners, he also helped to unscramble a series of disputes which had festered between the coal owners and the University. Sir Clifford Cory, another member of the Club, also became the first president of the School of Mines, which subsequently became Glamorgan Technical College, then Glamorgan Polytechnic, then the Polytechnic of Wales, the University of Glamorgan, and now the University of South Wales.

During the 1890-91 rugby season Ingledew formed a fine new partnership at half-back for Cardiff RFC with Ralph Sweet-Escott, another scion of a great sporting family who were members of the Club. Ralph was the third son of the Rev. William Sweet-Escott, the Rector of Penarth, and after attending King Henry VIII School in Coventry and Peterhouse College in Cambridge, he moved to Cardiff and trained as an architect under John Price Jones. There must have been plenty of chatter about sport as he worked under the leading figure in the world of Cardiff CC and Glamorgan CCC. Ralph himself played for the county XI, but like Ingledew, it was at rugby where he excelled and, besides playing together for the Barbarians, the pair were chosen by the Welsh selectors for the 1891 Home Nations Championship, against Scotland. Wales lost heavily by a penalty goal and six tries and the upshot was that the pair of Cardiff half-backs were dropped.

Though the Scotland game was the third and final appearance for Hugh Ingledew in Welsh colours, Ralph Sweet-Escott played again in 1894 and 1895, both times against Ireland. These were very exciting times for the young architect as, in November 1894, he was best man at the marriage of his elder brother Sidney to Ethel Brain, the eldest daughter of Samuel Arthur Brain. Educated at Trinity College, Oxford, Sidney joined the Cardiff and County Club in 1896 and subsequently became a leading figure in the family's brewing business. He also played rugby and cricket for Cardiff, as well as county cricket for Glamorgan CCC, often alongside Edward, the youngest member of the Sweet-Escott clan, who was also a Welsh hockey international.

Sidney Sweet-Escott's wedding had taken place at St. Augustine's, his father's church in Penarth, and the following week Ralph and Sidney's sister also married into the brewing trade as she tied the knot with Charles Hancock, the brother of Frank Hancock, one of their rugby-playing colleagues. The links between these two great Cardiff brewing families stood Cardiff in good stead in future years as hordes of people swarmed into the town to watch and celebrate great sporting events, as well as taking part in great civic events.

Sidney Sweet-Escott rose to a prominent position in Brains Brewery and was a familiar and popular face at the Club, but his younger brother Ralph died working in South Africa in 1907. Ralph had met much success on the rugby and cricketing fields of Cardiff and further afield, besides being a popular member of Cardiff Tennis Club and playing on the courts adjacent to the Club, but he chose to emigrate and work in the gold mining industry in the Cape. By the autumn of 1907 he was working at Knights Deep Gold Mine where he subsequently contracted enteric fever and died in mid-November.

The Sweet-Escotts and Ingledews represented the growing power of Cardiff RFC within Welsh rugby, as well as being at the heart of the world of law and brewing, as Cardiff's economy continued to boom during the first decade of the 20th century. Together with their many friends and business acquaintances in the Club, they were delighted in 1905 when the Welsh Rugby Union allocated their inaugural international match against the New Zealand All Blacks to the Arms Park.

Not only did the choice, in preference to the St. Helen's ground in Swansea, represent the way that Cardiff was now the powerhouse of Welsh rugby, it reflected the ground improvements which had taken place at the Arms Park. With the full support of the Bute Estate, a large grandstand had been built at a cost of more than £2,000 during 1900-01. It was funded jointly by the Welsh Rugby Union and the town's rugby club, and the capacity of the ground rose to 35,000. The Bute Estate, although primarily wishing to promote amateur recreation on the Arms Park, realised the massive money-spinner presented by rugby internationals, and this explained why a high wall had been built along Westgate Street, helping to enclose the ground and facilitate the collection of entrance money through the gates.

The Club was inside this perimeter wall and, given its impressive façade, it became an integral element of this new fortress of sport and Welshness, celebrated with great joy and pride on December 16th, 1905, when Wales

Glamorgan playing a match at the Arms Park, with its lavish pavilion, during the early 1900s. The rugby ground can be seen in the distance.

defeated the mighty All Blacks. Much has been written about the historic win over New Zealand, with historians regarding the victory as a seminal moment in the history of Welsh sport, and the emergence of Wales as a modern nation.

This tour in 1905 by the New Zealand rugby players to Europe and North America has gone down in sporting history as legendary. The Kiwis scored no fewer than 976 points and conceded just 59. They won 34 out of their 35 matches with their sole defeat coming in front of 47,000 jubilant Welsh supporters at the Arms Park. The week during the lead-up to the game had seen the Welsh side and officials based at the Queen's Hotel, with the All Blacks arriving on the Thursday before the match after trouncing Yorkshire by 40-0 and being mobbed by a frenzied throng of jubilant supporters as they walked from Cardiff General Station to the Queen's. The night before the game there was also plenty of chatter in the Club about what might happen the following day, though not even the most ardent or patriotic supporter of Welsh rugby could have predicted the events that unfolded.

After the All Blacks had run on to the ground to much applause, they performed their legendary 'haka' before the crowd, led by Dr. Teddy Morgan, the Welsh winger and medic from Swansea, sang the Welsh national anthem in an attempt to reduce the perceived psychological advantage of the Maori war-dance before the kick-off at 2.30pm. It was also the first time a national anthem had been sung before a sporting fixture. Teddy was soon in the action, sprinting some 25 yards down the wing to score a try out on the far left at the Westgate Street End. The roar of emotion and cheers of delight from the crowd were so loud that a carthorse bolted down Westgate Street and galloped straight past the front door of the Club.

Wales went into half-time with a 3-0 advantage but the tourists were far from happy, believing that the refereeing of Scotsman John Dallas was poor, especially in the scrums, and that in addition to not keeping up with play, he had blown his whistle too early for the break. Throughout the interval further Welsh hymns and anthems were sung as the crowd did their bit to maintain Welsh ascendancy. The second half proved to be tough and uncompromisingly physical as, fuelled by anger, a wave of All Black attacks took place before, with ten minutes remaining, Bob Deans almost reached the Welsh try-line before being tackled just short by Morgan and Rhys Gabe (though for years New Zealanders have claimed Deans placed the ball over the try-line). After some heroic defence by the Welsh side during the closing minutes, the final whistle signalled a totemic victory for the men in red.

While the victorious Welsh players were taken in a pair of four-in-hand carriages for dinner at The Esplanade Hotel in Penarth, the victory celebrations in the Club more than matched those of a couple of months earlier when Cardiff gained city status. There are no records of how many pints of beer, glasses of spirits, or bottles of fine wine and champagne, were drunk that night in the Club, but there was no doubting that, along with the people in all of the other watering holes in the new city, they had plenty to celebrate.

Rugby had now become a mass spectator sport in south Wales, with the New Zealanders' share of the record £2,650 gate receipts being much, much more than the £500 they had been guaranteed by the Welsh Rugby Union. Perhaps of most importance, though, was the fact that both Wales and Cardiff had come of age because of what happened during 1905 with events at the Arms Park on that heady day in December being the crowning achievement, and the Club now firmly established as a key element in every aspect of the work and play of the newly-designated city.

Teddy Morgan.

The love affair with rugby and the Arms Park

For modern-day members of the Club, the Saturdays, Sundays or even Friday nights when Cardiff plays host to matches in the Six Nations Rugby Championship, sees the Clubhouse come alive with young and old, enjoying the fulsome hospitality before the match, and then either strolling across to their seats in the nearby stadium, or taking a pew upstairs on the first floor to enjoy the television broadcast on a big screen, before re-assembling downstairs in the dining room to enjoy post-match hospitality and to chew over events from the match and any refereeing decisions which may have impacted on the result.

Apart from the use of a digital projector to display the television broadcasts, little has really changed over the years on international days at the Club, with members proud to use the facilities just a drop-kick away from the dead-ball line and to share in Wales' triumphs. The victory over the 1905 All Blacks was a landmark occasion, not least in the history of the city of Cardiff, but in Welsh history as a whole. As Martin Johnes wrote during 2005 in his history of sport in Wales, the match 'laid down the basis of the relationship between rugby union and Welsh national identity. The leaders of the new south Wales liked to see the popularity of rugby as a symbol of the unity of a society and its people. To them, the Welsh successes at the game were further evidence of the achievements and virtues of a nation: a metaphor for the wealth and confidence that industrialisation had brought.'

The same could certainly be said in the 21st century, with the corridors and rooms of the Cardiff and County Club full of members ready to toast the achievements of the Welsh team.

10

The Great War

"The Club believes that a measure of formality is prized, but friendship is of the greatest importance – that's why the Club is so popular."

The first decade of the 20th century was part of what sporting and social historians have called 'the golden era'. From Cardiff's point of view, 1905 had been the pinnacle of these years, with Cardiff's city status and Wales' rugby success over New Zealand being met inside the Club by great celebrations. Further success by the Welsh rugby team was met with further acclaim, as were the successes of the Club's members in other sporting events.

Golf, in particular, had become a popular pastime, with the courses at Whitchurch, Radyr and Llanishen all being venues within the city's boundary where an active group of Club members regularly played and challenged each other. It wasn't just the golf courses within Cardiff where members played, or held influence: Arthur Ingledew, who spent many long hours working alongside his brother Hugh in the family's solicitors' practice, acted as secretary of the Glamorganshire Golf Club which had come into being in Lower Penarth after the Earl of Plymouth had created a nine-hole course on his land. The course was subsequently expanded to 18 holes in 1896 and, from the following year, hosted the Welsh Amateur Championships. In the following years, the Glamorganshire Club attracted the interest as well as the participation of many members of the Club, with Hugh and Arthur Ingledew revelling in having a round with their sporting and business friends. The Sweet-Escotts, whose family home was nearby, were others to actively take part in these golfing events and, from 1901, their friends in the Barbarians rugby club also participated in what became the annual 'golf frolic' during their Easter tour of south Wales.

There was still a very strong hunting set in the Club during the early 1900s, with plenty of talk in the Club's rooms about equine events, as well

A group of members of the Glamorgan Hunt in 1934. Left to right – The Mackintosh of Mackintosh, Bob (R.H.) Williams, Mrs. Diddy (R.H.) Williams, Mr. Grant (huntsman), Sir Cennydd Traherne and Mr. Chapple (head groom).

as the annual point-to-point, overseen by Samuel Gibbon of Cowbridge who, between 1886 and 1897, had been Master of Foxhounds of the Glamorganshire Hunt. By the turn of the century, a number of other Club members were active participants with the Glamorgan Hunt including the Brain brothers, as well as the sons of Colonel Homfray of Penlline Castle, and Colonel Richard Bassett, who succeeded Gibbon as Master of Foxhounds before handing over in 1906 to The Mackintosh of Mackintosh who had regularly ridden in the Club's point-to-point.

Having joined the Club in 1880, The Mackintosh lived at Cottrell Mansion at St. Nicholas and owned extensive areas of land in Roath where during the closing years of the 19th century, vast areas of housing were created on his land. As a keen lover of sport, he ensured that the leisure and recreational needs of the new residents were also catered for, and during 1890 Plasnewydd House was renamed the Mackintosh Institute and converted into a sports club. In many ways it mirrored the facilities which The Mackintosh had enjoyed at the Club, with the new club containing a library, reading room, billiard room and gymnasium, with the two acres of grounds around the house being used for tennis, croquet and cricket. The operation of these facilities and the Mackintosh Institute itself was overseen by a board of trustees under the chairmanship of Lewis Shirley, the agent to the Mackintosh Estate and another active member of the Cardiff and County Club.

For many members of the Club during the Edwardian era, the highlights of the sporting calendar remained the race days at the Ely racecourse, reached by train from Cardiff General or by pony and trap along Cowbridge Road. An especial favourite was the spring day when the course played host to the Welsh Grand National, and when many of the country's leading steeplechasers took part in a long-distance race, inaugurated in 1895, over the Cardiff course. With members of the Club being well connected in the

---

Temporary Members

During the First World War a number of military officers, who were billeted in south Wales, joined the Club on a temporary basis as this extract from the Club's membership register shows.

world of National Hunt racing, it duly led to many busy days, and long nights, at the Club.

There was therefore a sense of vibrancy in the corridors and lounges of the Club during the opening decade of the 20th century. Club membership had continued to grow, with the membership ranks being swelled by a host of young men, many of whom were sons and brothers of existing members who were thoroughly enjoying the Club's new facilities. Others were part of the new wave of men whose success in both the business world and in local sporting circles also saw them gravitate towards the lavish and grand premises in Westgate Street. Their cheery chat and jolly tales of derring-do enlivened many evenings in the Club's rooms and at the many balls and other social functions held during the early 20th century.

At first, 1914 looked like being no different but, as the summer months unfolded, the conversations took on a more serious air, first with talk about the ever-worsening situation in Ireland, and then with news of labour unrest and further strikes at docks in the UK. Confirmation of the death of Jack Brain on June 26th cast a dark shadow over the Club, but the mood became far more serious two days later following the assassination of Archduke Franz Ferdinand in Sarajevo. As events unfolded in Europe during July and early August, the Club's members, like others across the country, realised that Britain would soon be at war with Germany.

The formal announcement of war, on August 4th, came as no surprise to many Club members who were members of the Territorial Army or the Glamorgan Militia. Like other reservists, they had taken part in regular training exercises, in readiness for deployment to quell labour unrest in Wales or in case they were required in Ireland but, from mid-summer onwards, events in Eastern Europe and the escalating disputes between the various Empires and powers, meant that they were preparing for a journey across the English Channel and into France and Belgium with what became known as the British Expeditionary Force.

The members of the business community had also been on a war footing for several weeks, with troops having been posted in the docks and around other prominent buildings in the city, as talk of a German invasion filled the bars, cafes, restaurants and offices throughout south Wales. Colliery owners and others in the coal trade also received advice from the Admiralty that emergency orders for Welsh coal might soon be placed.

Following the declaration of war, the Great Western Railway began running an emergency timetable and many of the planned excursions over the August Bank Holiday period were cancelled as troop movement became the number one concern. A recruitment station was swiftly opened at Cardiff City Hall and, within a couple of hours, as in many other towns and cities, hundreds of men had enlisted. Four German vessels were seized at Cardiff Docks, and their crew taken to temporary internment camps. On August 5th, the first of a wave of military detachments marched through the main streets of the city, prior to boarding trains at Cardiff General to travel to Bordon Camp in Aldershot. 'The streets of Cardiff remained crowded up to a late hour at night,' reported the *Western Mail*, 'and before the trains steamed out, there were some affecting scenes as wives and sweethearts took farewell of the soldiers. Cheers echoed through the station and were taken up by the dense crowds who had congregated.'

Conversations between the gentlemen in the Club were now based around which regiments they were associated with, and which battalions their sons or other family members would soon be joining. The mood at first within the Club reflected the feelings of many in south Wales as it was believed the conflict would all be over by Christmas. But within a few months, as news came through of the death on the Western Front of sons, brothers and friends from the Club, there was a darker and more morbid realisation that going off to fight for King and Country was a far more serious and deadly matter.

August 5th, 1914 had also seen Lord Kitchener being sworn in as the Government's War Secretary, and the following day he made his now famous appeal for rallying a further 100,000 men for the Army with the slogan that 'Your Country Needs You'. Kitchener believed that overwhelming manpower was the key to success, and he duly set about looking for ways to encourage men of all classes to join. It was an approach that was in direct contrast to British military tradition which, for hundreds of years, had relied on professional and well-trained soldiers rather than conscripted men.

Several leading members of the Cardiff and County Club responded by playing a prominent part in the initial phase of recruiting, including John Nicholl of Merthyr Mawr House. The Old Etonian was the son of John Cole Nicholl, the former Tory MP for Cardiff, and was prominent in a number of recruitment and training initiatives in the Vale of Glamorgan and at Lavernock, where various Territorial groups took part in regular summer camps. Others subsequently played a leading hand in the recruitment drive, especially with the so-called 'pals battalions', or 'lads battalions' formed in response to Lord Kitchener's rousing appeals. These comprised men who had enlisted together in local recruiting drives, with the promise that they would be able to serve alongside their friends, neighbours, work colleagues and team-mates, rather than being arbitrarily allocated to other battalions.

On November 2nd, 1914, the National Executive Committee called on the Lord Mayor of Cardiff, Alderman J. T. Richards, to raise a battalion from the Cardiff area, tapping into the wave of patriotism with the new battalion wearing the arms of the city. A fortnight later, the War Office authorised the creation of the Cardiff City Battalion and, four days after that, the recruiting office was opened, with Captain Frank Hill Gaskell appointed in command of the new brigade. He had been recovering at his home in Llanishen, the popular suburb to the north of Cardiff, having been wounded in the face during early skirmishes when serving with the Third Battalion of the Welsh Regiment who were part of the British Expeditionary Force in France.

The Gaskells

Brains Brewery did not dominate the drinking habits of Cardiffians during the years leading up to the Great War as a number of other breweries sought a decent share of the market. The greatest competition came from Hancock's, one of the leading breweries in the West Country which, in 1883, established new premises in Cardiff, known as The Bute Dock Brewery.

The manager of the new enterprise was a Devonian called Joseph Gaskell who had moved to Cardiff from Newton Abbot to set up business

Joseph Gaskell (middle row, second left), sitting with his wife Elsie and other family members during the 1920s.

initially as a cider agent. Sales of Hancock's beers soon increased and the Gaskell household moved from Windsor Road in Splott to more spacious – and upmarket – family accommodation in Penarth, where Joseph and his Yorkshire-born wife Emily raised their family.

With Hancock's trade continuing to thrive, the Gaskells subsequently moved to live at The Coldra in Newport before acquiring New House in Llanishen on the slopes of Caerphilly Mountain, where Joseph and his family could enjoy commanding views of the Cardiff neighbourhood as well as, on a clear day, a view of the coastline in north Somerset and north Devon.

With his rising status, Joseph joined the Cardiff and County Club in 1903 and he subsequently became a regular face in and around the Clubhouse, besides helping to secure a supply of the Hancock's brew for members to enjoy. Despite being a business rival of Jack and Sam Brain, the two brothers enjoyed Joseph's company in the Club, as did several other members of staff from The Old Brewery, proving that any business rivalries were left outside the door of the Clubhouse in Westgate Street and, like so many others, once inside the premises they were all good friends.

Frank was Gaskell's second son, born in Penarth in 1878 and, after attending Llandaff Cathedral School and Malvern College, Frank served with the Volunteer Company of the Welch Regiment in the Boer War, before returning to Cardiff to complete his studies at University College and qualify as a barrister. After graduating, he followed his father by joining the Cardiff and County Club, before becoming a member of the South Wales Circuit, as well as serving as a county councillor for the Adamsdown Ward from 1906 to 1908.

By this time, Frank Gaskell had moved with his wife Violet – the sister of one of his old school friends from Llandaff – to Boscobel, a detached house in Station Road in Llanishen. Through his membership of the village's cricket and tennis club, Frank also rubbed shoulders with other leading residents of the delightful suburb, plus other friends from the Club, including Ivor Downing, an Oxford- educated solicitor whose family had been long-standing residents of Llanishen and who subsequently married Edith Gaskell, Frank's younger sister.

In 1900 Frank had joined the 3rd Volunteer Battalion, Welch Regiment, and served in the Boer War until 1902, where he was awarded the Queen's South Africa medal for his service abroad. He remained in the Volunteer Battalion and, following the outbreak of war in 1914, served in France with the 2nd Battalion, Welch Regiment, as part of the British

Frank Gaskell was portrayed in this Staniforth cartoon published in the Western Mail on September 5th, 1914.

Expeditionary Force but after a couple of days in the front line, he was shot in the jaw and returned home to recuperate.

On September 19th, 1914, David Lloyd George, in a rousing speech at the Queen's Hall, London, had called for the formation of a distinct Welsh division within the British Army. Three weeks later, the War Office formally agreed to the creation of the Welsh Army Corps and, after meetings involving the Lord Mayor of Cardiff, approval was given for the 16th Welch Regiment (the Cardiff City Battalion) with Frank Gaskell being invited to oversee the recruitment process.

By now, the war was into its 16th week and the first rush of volunteers had long subsided, but the military authorities believed that Frank's political, military, professional and social background made him the ideal man to muster around 1,000 men. He duly oversaw a series of public meetings, concerts by military bands, grand military demonstrations, and appeals at soccer matches, places of work and music halls, and he used his contacts from the Club and the legal community to drum up further support.

Enrolment took place at the Custom House in Cardiff and, after eight weeks, enough men had enlisted, as a host of sportsmen from

> across the city and its suburbs – especially from Cardiff RFC – dutifully signed up in a process of recruitment mirrored across other towns and cities in Wales as the sporting pals all joined up together. Sadly, many were to die together on the bloody battlefields of the Somme with the Battalion suffering more than 450 casualties, including over 150 deaths, at Mametz Wood alone in 1916. This heavy loss of life was a grievous blow for the city, as across its homes – both rich and poor – there was the loss of sons and brothers. For Joseph Gaskell bad news had arrived earlier at New House as Frank was killed after a bullet had struck his ammunition pouch, which ignited causing him to sustain a fatal wound to his stomach.
>
> Shortly after his death, the *Western Mail* published extracts from a letter sent by a colleague back home to his mother in Cardiff. It told of Frank's death and the high regard in which he was held – 'He always had a cheery word when he used to go through the lines in the night. Only half an hour before he was wounded he passed us and said, "I hope you are giving them a good peppering lads!" He had any amount of pluck and would never let a man go where he would not go himself.'

Like so many families across Cardiff and south Wales, the next four years saw almost every gentleman associated with the Club lose a son, brother, relative or friend in the conflict which raged until the Armistice in November 1918. Sir John Courtis, the Lord Mayor of Cardiff in 1911-12 and the man whose brickworks on the northern outskirts had literally supplied the building blocks needed for the city's housing boom, lost his son John Harold. The Australian-born entrepreneur lived at Fairwater Croft in Llandaff, with John junior attending Llandaff Cathedral School and later Repton, before entering the Royal Military College at Sandhurst. He joined the Oxfordshire and Buckinghamshire Light Infantry in 1908 and rose to the rank of captain. But in November 1915, when carrying despatches near Basra in Iraq, the 27-year-old was killed by a sniper's bullet.

Dr. Henry Ensor, the senior ophthalmic surgeon at Cardiff Infirmary, also lost his son, John. The Ensor family lived at The Hollies in Llanishen, with John's grandfather, Thomas Henry Ensor, having been one of the founder members of the Club, and a leading figure in the legal and political world of the town with his duties including chairing the Conservative Club. Henry had read medicine at Guy's Hospital in London before training at a hospital in Birmingham. In 1887 he returned to Cardiff to work at the Infirmary, joined the Club and settled in Llanishen where John was born in 1895.

After attending Llandaff Cathedral School and Epsom College, John looked destined to follow his father into the medical profession, having secured a scholarship to read medicine at the University College of South Wales in Cardiff. Shortly after starting his studies, John joined the university's military training corps and, in 1915, accepted a commission with the Welch Regiment. He went to the Western Front where, in late November 1917, he was wounded during skirmishes at Bourlon Wood in the Battle of Cambrai. He never recovered and died of his wounds two days later.

Major Evelyn Orlebar, one of the leading lights in the Club's polo team, also lost his only son, Robert, who was killed in January 1915. Educated at Llandaff Cathedral School and Cheltenham College, Robert followed his father into military service after attending the Royal Military College in Sandhurst and being gazetted to the 2nd Battalion, the Middlesex Regiment (Duke of Cambridge's Own) in September 1913. After time in Malta, his battalion was sent to France the following November as much-needed reinforcements for the British Expeditionary Force. But early in the New Year, the 20-year-old was shot and killed by a sniper while off-duty in the trenches at Rue de Bassiere near Neuve Chapelle.

Walter Shirley was another member of the Club to lose his son. The well-known solicitor and his wife received the awful news in early June 1917 that their son Archibald Vincent had been killed in a flying accident near the small Belgian town of Roulers. Born at the family's home in Marine Parade in Penarth in 1887, Archie had been named by his devoted parents after his paternal grandfather – Archibald Hood, the owner of the Glamorgan Coal Company and one of the most influential figures in the industrial world of the Rhondda. Educated at Llandaff Cathedral School, Rugby School and Exeter College, Oxford, Archie was a law graduate who had joined his father's practice in High Street and looked set to follow in his illustrious footsteps. But it was not to be as, after joining the Royal Flying Corps and passing the training course, he was stationed in Belgium. On June 8th, 1917 he was killed flying a Sopwith Pup which collided with another biplane belonging to 66 Squadron.

One of the first to offer Walter his condolences was his good friend and polo partner Lionel Lindsay. The latter's family were also touched by sadness, but few met with quite the series of tragedies during the Great War that befell his elder brother Morgan Lindsay, another prominent member of the Club who had served in Northern France and Belgium from early August 1914 with the British Expeditionary Force.

11

Footballer, Father and Friend

"It's been regarded as the last bastion of the Raj this side of Calcutta!"

The membership lists of the Cardiff and County Club have always contained captains of industry, leaders of local society and politics, as well as prominent figures from practically every walk of life in the town and city. Also on the rolls have been famous sportsmen, with Welsh rugby players, jockeys, golfers as well as county and Test match cricketers, besides others who have won Blues at Oxbridge or medals at prestigious international events. From 1887, the club could also boast on its members' list the first Welshman to play in an FA Cup Final. However, the story of Morgan Lindsay, the second son of Lieutenant-Colonel Henry Gore Lindsay and the daughter of Lord Tredegar, was not one full of glittering sporting success, as during the First World War, he suffered perhaps more personal tragedy than any other man who frequented the corridors and bars of the Clubhouse in Westgate Street.

Born at Tredegar Park in February 1857, Henry Edzell Morgan Lindsay had grown up at the family's home, Woodlands, in Leckwith before following his father into the military by joining the Royal Engineers. Morgan also shared his father's love of sport and, besides playing cricket and riding with the local hunt, he played football for the Royal Engineers and in 1878 appeared for the military team in the final of the FA Cup, held at The Oval. It ended in a 3-1 defeat for the Wanderers but Lindsay entered the record books as being the first Welshman to appear in an FA Cup Final.

Colonel Morgan Lindsay.

During the 1880s he served in both South and Central Africa, including in the latter a series of skirmishes against the Mahdist Sudanese, largely comprising men from the Hadendoa tribe who were known to the British troops as 'Fuzzy Wuzzies' because of their unique hairstyle. He later returned to south Wales, where he continued to play cricket for various gentlemen's teams besides riding with the Glamorgan Hunt and taking an active part in the Club's many social functions. Indeed, it was at the grand balls where his romance with Ellen Thomas developed. She was the daughter of George Thomas of The Heath, where horse racing had taken place during the Victorian era. The pair met at a local point-to-point and they soon became an item before getting married on July 24th, 1889. Within a couple of years, they had made their home at Ystrad Fawr, a spacious house belonging to the Thomas family at Ystrad Mynach and, over the course of the following eight years, the couple had three sons and two daughters.

Morgan also became active in local politics, joining the Caerphilly Urban District Council in 1892, before being elected on to the Glamorgan County Council, and rising to the position of Chairman of Caerphilly UDC in 1914. In 1900 he was also nominated as Conservative candidate for the East Glamorgan parliamentary constituency but, by the time of the general election, he was on active service again, this time with the Monmouthshire Royal Engineers in the Boer War. Ellen duly oversaw his election campaign against Alfred Thomas, the chairman of the Welsh Liberal Parliamentary Party, who had held the seat since 1885, and went on to again secure a majority in the election.

After returning from the Cape, Morgan continued to serve on Caerphilly UDC, besides being a keen and generous supporter of cricket and football, as well as riding in point-to-points and breeding thoroughbreds from his stables at Ystrad Fawr. One of the horses he bred was Dean Swift which came third in the 1897 Welsh Grand National, run on April 19th, at the Cardiff Racecourse,

Morgan Lindsay (right) on his return from the Boer War in 1901.

and ridden by his brother Walter, much to the delight of the Lindsay clan and his friends at both the Club and in the Glamorgan Hunt.

Given his long and distinguished military record, it was fitting that Morgan should join up again with the Glamorganshire Yeomanry at the outbreak of the Great War in August 1914, and he briefly spent time in France with the British Expeditionary Force before retreating to the UK following the Battle of Mons. He also encouraged his three sons, all of whom had been educated at Wellington College, to join up. His sons had inherited their father's love of sport with all three winning a place in the Wellington 1st XI. George and Claud had also become cadets at the Royal Military Academy at Woolwich, with their mother and father sitting proudly in the audience when George passed out from one of the world's foremost military colleges in 1911. Little could the proud parents have known that, within half a dozen years, their eldest son would be killed in a tragic accident while testing planes near Bristol.

Like his father, George had been among the first wave of British troops to depart for foreign fields in August 1914, sailing to France with

Mrs. Ellen Lindsay and three of her children. Both boys were killed during the First World War.

the Royal Field Artillery. In November he was wounded during the Battle of Ypres and briefly returned to his family home to recover, before returning from abroad and commanding a battery at Salonika during 1915. In November 1916 George joined the Royal Flying Corps and, after flying a series of successful sorties over enemy lines in France and Belgium, was deployed back in Britain during the summer of 1917 to test a series of newly-built and repaired aircraft from the Rolls-Royce factory near Filton Airport in Bristol. With events on the Western Front poised to move into a decisive stage, his new duties for the war effort were important in ensuring a decent supply of fit-for-purpose planes. Tragically his work – ironically well away from enemy fire – was to cost him his life, as he was killed, together with an air mechanic who was aboard the plane, when it crashed near Chipping Sodbury on June 25th, 1917.

Morgan Lindsay at the races in Cardiff during the 1920s.

It was a grievous blow for Lindsay and his wife but, within nine months, the Lindsay family were dealt two further blows as two other sons, Archie and Claud, were killed within a week of one another on the Western Front. Archie had followed his father in a career in the Royal Engineers and was serving as a Lieutenant in the Seventh Royal Monmouthshire Brigade when it was caught up in skirmishes in the Pas-de-Calais area, with Archie losing his life on March 26th.

Later that week, news arrived at Ystrad Mynach House that Claud had also been killed in action on the Western Front. He had enjoyed a glittering career at both Wellington College as well as RMA Woolwich and by the summer of 1918, the 26-year-old had risen to the rank of Acting Major in the 33rd Battery of the Royal Field Artillery. In June 1915, during a short spell back at home, Claud had married Dorothy Forde and, in early March 1918, before the advance on the Western Front, he had briefly returned home to see his wife and family. Tragically, it was the last they saw of him as he was killed in action on Easter Day, March 31st, during an Allied offensive on the Somme.

The tragic loss of their two sons within a week of each other was tempered by news, a few weeks later, that Dorothy was pregnant and on

November 16th, 1918 – like so many other war widows – she gave birth to a boy, named George Morgan Thomas who, like others born at that time, never knew his father. The birth of the little grandson was a modicum of comfort for Lindsay and his wife, and one can only wonder at how they coped with such personal tragedy. As a devout Christian, Lindsay took great solace in his faith, believing that his sons had died for the greater good and, during the years that followed, he threw himself more and more into the world of sport, possibly as a way of forgetting what had happened during those grim and dark days during the war.

With the support of his wife, and his many friends at the Club, Morgan Lindsay obtained a training permit from the Jockey Club and became a familiar face at races across Wales and the West of England. The highlights of his training career undoubtedly came during the 1920s when his horse, Miss Balscadden, won the Welsh Grand National in 1926 and 1928 at Ely Racecourse. The mare was owned by Sir David Llewellyn, the father of Harry Llewellyn of Foxhunter fame, and was one of several horses owned by Sir David which were in the care of Morgan Lindsay at his Ystrad Mynach stables.

The star of his stables, though, was another steeplechaser called Ego. A winner of the National Hunt Chase at the Cheltenham Festival, it also finished third in the Welsh Grand National run at the Ely course in April 1935. It was the last big winner trained by Morgan Lindsay. A few months later he was taken ill and, for the next few weeks enjoyed poor health, before passing away at his home on November 1st.

The Clays of Piercefield Park

If Morgan Lindsay was the first member of the Club to train a winner of the Welsh Grand National, Johnnie Clay – the Glamorgan and England cricketer – was the first to have a prestigious steeplechase run in his memory each year at a major racecourse.

Johnnie was a member of the Clay family who lived at Piercefield Park to the north of Chepstow, with the family's home and extensive grounds overlooking the Wye Valley. His grandfather was a banker and brewer from Burton-on-Trent, who had acquired the property in 1861 after moving to work in the thriving industries of south Wales. His father became a leading ship owner at Cardiff Docks and, by the late 19th century, the first of a long line of members of the sporting family were regular faces in the dining rooms and bar of the Club.

Educated at Winchester, Johnnie had mixed riding and hunting in the winter months with playing cricket in the summer. Initially a tearaway fast bowler, he had played in Minor County cricket for Monmouthshire, before making his Glamorgan debut in 1921. After a series of injuries, he switched to spin bowling and became one of Glamorgan's finest bowlers. In 1935 he also won an England cap, and appeared in the Fifth Test against the South Africans. In fact, he might have won many more caps but turned down the opportunities, urging the Test selectors to play younger and fitter men.

Johnnie Clay during the 1920s.

Indeed, in 1938 he spurned the opportunity to play against Don Bradman and the rest of the Australian side, having sustained a knee injury when demonstrating to his children the steeplechase prowess of Golden Miller. He was leaping over a series of chairs at his home in the Vale of Glamorgan to show how the famous horse had won the Cheltenham Gold Cup when, unlike the horse, Johnnie took a tumble. The knee injury caused him to miss several matches.

By the time Bradman and the Australian cricketers returned for the 1948 Ashes series, Johnnie had become an England selector. During the second half of the summer, the 50-year-old also played a major hand in Glamorgan securing the County Championship title for the first time in their history. After playing his final county match the following year, against Yorkshire at Newport, he maintained his involvement in the sporting world by acting as secretary of the Glamorgan Hunt, and serving as both a steward and a director of Chepstow Racecourse, which had been laid out at Piercefield Park where Clay had his home near Chepstow during the late 1920s.

In 1925 a group of ten businessmen and leading figures from the social world of south Wales, including Courtenay Morgan, the 1st Viscount Tredegar, had formed a company to purchase Piercefield House, and lay out a racecourse in its grounds. The first meeting duly took place on August 6th, 1926, but the course nearly closed down due to cash problems, and it was only after a large bank loan guaranteed by the directors and the Clay family, that racing continued, with the first jumps meeting taking place the following March.

For almost 50 years Johnnie and his sons owned a string of fine steeplechasers and, in 1972, he won the Welsh Grand National with Charlie H, jointly owned with Susan Williams, the daughter of the Master of the Glamorgan Hunt. Trained by Bob Turnell at Lambourn, Charlie H had been owned originally by Jim Joel, but he had ordered the sale of the bay gelding as he didn't believe that he was up to much. Johnnie and Susan thought otherwise and their judgement was justified as the ten-year-old was given an enterprising ride by jockey Johnny Haine on heavy ground to win the nine-runner race by two and a half lengths from Fair Vulgan, who was receiving a stone from the winner, after Charlie H had taken the lead at the second last, before going on to win with something to spare.

Johnnie Clay was slightly taken aback by the victory, as he thought Charlie H's weight of 11 stone 3 pounds would be too much in heavy ground around the stiff track. But his friends at the Club and among many hunts across south Wales, had broad smiles on their faces after backing the 11-2 winner. Following Johnnie's death in August 1973, it was very fitting that a long-distance steeplechase was established and run annually at the Chepstow racecourse in his memory.

Johnnie Clay (standing, back left) with other members of the Glamorgan Hunt, and the Club. Also standing at the rear are Colonel Pritchard, Charles Cory and 'Tip' Williams with, at the front, Bill Williams, Bob (R.H.) Williams, Mrs. Evie Homfray, Colonel H.R. Homfray, Mrs. Diddy (R.H.) Williams, John Cory and Mrs. Mary Williams.

12

The Roaring Twenties

"A haven of Edwardian splendour where you can go and talk to anyone without having been introduced beforehand."

'It's Over' – so proclaimed the newspaper billboards on Monday, November 11th, 1918, following the signing of the Armistice, and the cessation of hostilities at 11am that morning. Shortly after lunch, all work ceased in much of Cardiff, like other towns and cities across the UK, as the news was confirmed. Lessons stopped at the schools, church bells rang out and work sirens were hooted in celebration and within an hour or so the streets were full of joyous men, women and children, many of whom were waving flags in sheer delight that the bloody conflict was, at long last, over. In the Cardiff and County Club, great celebrations also took place and during that long, long night, the management and Club members alike raised a glass, or two or three, to their friends and acquaintances who had died.

Over the course of the next few days, a number of other events took place in the vicinity of the Club, as a thanksgiving week was organised by civic authorities in Cardiff. The festivities included a special rugby match at the Arms Park between Cardiff and Tredegar, and among the decent-sized crowd were more than 500 American sailors who had been given special permission to attend after the military authorities, who had taken over the recreation ground and the large pavilion, had given their blessing to the game taking place. The pavilion, as well as the Cardiff Racquets and Fives Club, had also been used as military hospitals, with wounded troops being treated before being moved to convalescent homes, located in some of the other large country houses elsewhere within the city, its suburbs and the Vale of Glamorgan.

The rugby match between Cardiff and Tredegar was the first tangible sign that life would soon, albeit very slowly, be getting back to normal. December 1918 and the early months of 1919 saw the city welcome back

An aerial view of the Arms Park and central Cardiff taken during the 1920s.

its brave men, as well as many women, who had done sterling service in a variety of capacities for the war effort. 1919 also saw Charles Berkeley briefly return as Secretary and Manager of the Club, following the retirement of William Herbison. Like many Cardiffians, Charles had been saddened to see so many of his friends and acquaintances lose their lives in the war, but he was heartened to see his old pals return from the Western Front, and it was with a mixture of sadness and pleasure that he listened to them recount their deeds while sitting again in the comfortable surroundings of the Club.

Trade was soon back on the up as life returned to normal during the 1920s and the positive mood within the country during the next few years was reflected in the world of sport as Glamorgan CCC secured, at long last, first-class status, whilst Cardiff City met with much success in the First Division of the Football League, as well as winning the FA Cup in 1927. Both

The Roaring Twenties

The 'Fire Engine Station' opposite the Clubhouse on Westgate Street, as seen in 1913.

the football and cricket clubs had the financial support of many members of the Club and the Exchange Club.

Various members of the Cardiff and County Club played an important role in nurturing the development of sport in the city during the 1920s. Foremost among these was Sir Sidney Byass of Llandough Castle, the owner of the Margam Abbey Steelworks and a member of the Cardiff and County Club since 1910. Born in 1862 in Surrey and educated at Radley College, he had been in the school's cricket team from a young age before studying at University College, Oxford. After the Great War, Sir Sidney helped several sporting organisations, including Aberavon RFC, in addition to underwriting Glamorgan County Cricket Club's application to the MCC for first-class status. His generous loan of £1,000 allowed a series of guarantees to be given to other English counties regarding fixtures with the Welsh county and, as a result, the MCC endorsed the bid and recognised the club as first-class for the 1921 season.

Their rise from the third-class ranks symbolised the feel-good factor which permeated so many aspects of life in south Wales during the 1920s as well as further illustrating the close links between the business and sporting community, which the Club had played no small part in nurturing. The opening first-class fixture in May 1921 was met with great delight at the Arms Park, and by the enthusiastic onlookers on the Club's balcony, as members were able to share in the success of Glamorgan, led by Cardiff dentist Norman Riches, in defeating Sussex.

Ernest Prosser

Ernest Prosser was typical of the local men who, through hard work and diligence, rose during the late Victorian and Edwardian era from fairly modest means to a prominent position in the public life of Cardiff. He had started working at the age of 14 as a junior clerk with the Rhymney Railway and, by the time of his death in 1933, he was general manager of the company, as well as its associates, the Cardiff Railway and the Taff Vale Railway, besides owning Parc Cefn Onn and many acres of land to the north of the city. A member of the Club since 1916, he had been awarded the CBE in 1920, largely in recognition of his outstanding work during the Great War as the deputy director of Railway Movements in the War Office.

His father Thomas had moved to Cardiff from his native Radnorshire and, after starting work as a carpenter, he created a thriving building company which erected many properties in the expanding suburb of Cathays. Indeed, it was in Woodville Road during February 1867 that Ernest was born, just a short distance from the railway line heading north to Caerphilly and the engine works where his fascination with steam locomotives began. He must have inherited some of his father's business nous because, within a handful of years of joining the Rhymney Railway as a junior clerk, he had been promoted to an administrative position and in 1905 became the company's general manager.

He retired in 1922 as the Rhymney Railway along with the Cardiff and Taff Vale companies merged with the Great Western Railway. By this time he was also a director of the Cardiff Gas Light and Coke Company, as well as Crosswell's Brewery, besides being a city magistrate. Ernest was also a leading figure in the life of Llanishen, the suburb to the north of the city where he had moved with his wife Florence in 1891 to live at Fernbank House on Station Road, just a short walk from his company's railway station. Besides being a leading light at St. Isan's Church, he was, at various times, captain and secretary of the Llanishen Cricket and Tennis Club, and from 1906 was the President of Llanishen Athletic Club.

In 1909 he acquired Cwm Farm, an extensive area of meadow and woodland on the southern flank of Caerphilly Mountain, He leased some of the farmland to Llanishen Golf Club, allowing them to create an 18-hole course to replace their previous nine-hole version on land belonging to Godfrey Clark at Ty Mawr Farm near Lisvane. Indeed, the Llanishen club, formed in 1905, was mainly a resurrection of the earlier

Lisvane club, formed in 1900, but lapsed owing to difficulty of access by either road or rail. There was no chance of the same happening at Cwm, or Cefn Onn as it became known, as Prosser wielded his influence by authorising the opening of a railway halt to the south of the Caerphilly Mountain tunnel and adjacent to the new course.

As well as a glittering career in business, and a fine record of public service to the people of both Cardiff and Llanishen, the story of Ernest Prosser also contained much personal sadness. Tragically, his wife Florence died in April 1896 at the age of just 29, and just six weeks after the baptism of their only son Cecil. The young boy grew up with his father and grandmother in a quite substantial property called Heathfield in Fidlas Road before embarking on a military career with the Royal Garrison Artillery.

Cecil served with the Royal Garrison Artillery between 1916 and 1918 in the Great War, and rose to the rank of second lieutenant, but shortly afterwards he contracted tuberculosis and was invalided out of the active service. With no drug therapy available, many afflicted by the illness went away to a sanatorium to enjoy fresh air. But with Ernest having extended the park known as The Dingle near Cefn Onn Halt, he decided to create a summerhouse where Cecil could spend as much time as he wished, and it was to here that the young man was transported, either by horse-drawn carriage from Fidlas Road or by train from Llanishen. He enjoyed several years there, spending time in the small rooms or sitting on the veranda, but tragically Cecil's condition deteriorated in 1921 and Ernest had to oversee his transfer to a sanatorium in north Wales, where Cecil died in November 1922.

It was typical of the man that Ernest dedicated a window in St. Isan's Church in memory of his late wife and son, as well as endowing a scholarship at the Welsh National School of Medicine to promote research into tuberculosis. His many friends at the Club rallied around after the loss of his beloved son but, as contemporaries observed, Ernest was never the same man after the second tragedy in his private life. He died in October 1933 at his home in Fidlas Road and was buried in St. Isan's churchyard alongside his beloved wife and son.

Though there may have been a buoyant mood in the business and commercial world of Cardiff in the early 1920s, life in the offices of the Bute Estate was less cheerful as the immediate aftermath of the Great War, and the vagaries of the south Wales economy, placed great pressure on the family who, during the late 19th century, had been one of the wealthiest in the world. With ever-rising costs and a need to rationalise his property and investments, the Fourth Marquess and his financial advisers had to consider every conceivable way of generating revenue, even if it meant disposing of the Arms Park which they had painstakingly promoted as a recreational space for the townsfolk, but was now regarded as something of a potential millstone around his neck.

When the Third Marquess died in 1900, some of his extensive estate was subdivided, but the bulk of his property followed the title, and John Crichton-Stuart, at the age of 20, became the major shareholder in the Bute Docks and the owner of his late father's land and rights in Cardiff and within the coalfield. He had inherited a personal estate worth more than £1.1 million, but death duties, family charges and sizeable bequests to the younger children of the family meant that the Fourth Marquess was far less wealthy than his late father.

Claud Williams (right) was a well-known member of the Club during the years either side of the First World War. He is seen here in the unsaddling enclosure at Chepstow Racecourse in 1928 with (left to right): Joe Frazer, Maude Frazer, Mrs. Diddy (R.H.) Williams, and Lady Glanely.

If the 19th century had seen the Butes amass great wealth, the 20th century was a time for dispersal. Property in the valleys was put on the market as early as 1909, and the sale of the Bute collieries took place between 1915 and 1919. A large number of land sales also took place in Aberdare, Treorchy and Treherbert immediately after the Great War, and during the early 1920s many farms and other freehold property within the coalfield were sold. Attention then turned to Bute interests in Cardiff, with the Bute Docks being absorbed in 1922 by the Great Western Railway Company, thereby relieving the Estate of its greatest financial responsibility.

1922 also saw the Marquess contact the tenants of the Arms Park, and the other properties lining Westgate Street, including the Cardiff and County Club, regarding the possible building development. The Club had already begun negotiations to secure the freehold of the property from the Bute Estate, and at an Extraordinary General Meeting of the Club held on February 25th, decisions were taken to raise additional funds, through the release of 400 debentures at £25 each, each carrying an interest of 5% per annum. This raised sufficient capital to proceed and an agreement was reached with the Bute Estate on April 21st, with the Club's Trustees – Henry Lewis of Greenmeadow, Bob (R.H.) Williams of Bonvilston House and John Griffin Thomas of Llandaff – signing off the paperwork.

The Marquess also informed the Cardiff cricket and rugby clubs – who were the major tenants of the Arms Park – that their agreeable relationship with their landlord was likely to cease. He added that they had first opportunity of purchasing the sports ground, for a price of £30,000, excluding a strip of land adjoining Westgate Street running down from opposite The Angel Hotel to the Racquets and Fives Club where a series of tennis courts had been laid out, and which was being earmarked for building.

The news was a massive blow to the city's sportsmen, and also raised the spectre of an end to international rugby in Cardiff. It united Cardiff's cricketers and rugby players who for many years had operated independently but now realised that the only way to preserve their hallowed turf was to join forces and raise sufficient capital to pay the asking price set by the Marquess and his advisers. The spring of 1922 duly saw the cricket and rugby clubs merge as Cardiff Athletic Club came into being with one of their first acts being the appointment of Hugh Ingledew to lead their negotiations with the Marquess.

A six-man panel was assembled under the chairmanship of Clifford Cory – another leading member of the Club – to draw up a plan to raise £35,000 for the purchase of the park by the Athletic Club to ensure its continuing operation as a sporting venue. Their plan involved raising £20,000 in £5 ordinary shares and £10,000 through 5% debentures. The Welsh Rugby Union also helped out by putting up £4,000 in debentures and, in return for their generosity, the Union was given the right to stage international fixtures at the Arms Park for the next ten years.

Many members of the city's business community and members of the Club dipped into their pockets to purchase these shares and, given the sizeable early interest, the Cardiff Arms Park Company Limited, as the consortium was known, contacted the Marquess to set in motion the formal purchase of the ground at the full asking price of £30,000. A downswing in the economy slightly slowed their fundraising activities and, after a couple of generous donations from two Club members – Sir Henry Webb of Llwynarthen and Eric Insole of Ely Court, who each donated £500 – the remaining capital was raised, and the formal completion of the deal with the Bute Estate took place on June 25th, 1923.

The racquets and fives court, however, did not remain in operation. The years after the Great War had been difficult ones, and the loss of personnel meant that the Club never really got going again. The tennis courts alongside Westgate Street were also less tranquil in these days of the automobile and, while the courts were earmarked for building purposes, the hall itself was sold in January 1925 to Cardiff City Council for £5,000 for the creation of the offices of its Juvenile Employment Bureau. The name of the Club's immediate neighbour also changed to Jackson Hall, after John James Jackson, the director of education for Glamorgan, with the new premises being formally opened by the Lord Mayor on September 20th, 1926.

Another change had taken place some years earlier on the opposite side of Westgate Street with the construction of a large fire station and five-storey building during 1917. Within the Club itself, there were a number of changes of personnel. The Club's President, Colonel Henry Lewis, died in 1925. He had taken over as President in 1917 after the death of George Insole, and had supervised the resumption of the Club's affairs after the Great War as well as the campaign to recruit new members following the loss of so many in action. Lewis had been an enthusiastic member of the hunting and shooting community and, from his home at Greenmeadow in Tongwynlais, kept a pack of foxhounds for the Ystrad and Pentyrch Hunt. A Colonel in the First Devon Yeomanry, he also had a collection of parrots which he kept in his dining room and trained to utter, at the appropriate time, some quite rich vocabulary. Colonel Lewis also had a good thirst for port and according to Club folklore swiftly devoured at least half a bottle every time he called into the Club for lunch.

His successor as President was Colonel Herbert Richards Homfray, the son of founder member John Richards Homfray and another gentleman with an active interest in country sports. Born in 1864 at Penllyn Castle, he had served as Master of the Glamorgan Hunt from 1906 until 1914, and was a regular face at the Cardiff races as well as the plethora of point-to-point steeplechases across south Wales which ran during the spring and early summer. The Old Etonian also had a military career with the Irish Rifles and subsequently the Second Battalion of the Welch Regiment before being a director of the Taff Vale Railway and chairman of the Cowbridge District Council, as well as their Board of Guardians. Homfray served as President of the Club until his death in 1940.

The 1920s also saw several changes in the Club's administrative and managerial personnel. In 1919 Thomas Tyrell had succeeded Charles Berkeley as Secretary but he remained in post for only a year before handing over to Walter Allen. He too soon moved on with Arthur Brett taking over but, during his term of office, there were numerous complaints about the quality of the food, and he left in April 1926. Thomas Duncan was his replacement but, within three months of

Herbert Richards Homfray.

Members of the Glamorgan Hunt gather for a meet in the Vale of Glamorgan during the 1920s.

his appointment, a series of financial irregularities led to his temporary suspension. He was reinstated but a further misdemeanour in March 1927 resulted in his dismissal. His wife though remained for a further nine months as Stewardess, with Lionel Eustace Taylor stepping into the breach as Secretary. He had been a Club member since 1906 and offered an element of stability until a suitable replacement was found as combined Secretary and Steward. This came early in 1928 when Captain Samuel James Price and his wife were appointed. They held the job for ten years and saw the Club through the next phase of its evolution and the start of another long and bloody war with Germany.

The Clubhouse, photographed shortly after the First World War.

13

The King and Freddie

"A most welcoming and completely class-less place."

The inter-war era saw royal patronage of the Cardiff and County Club. His Royal Highness, Edward, Prince of Wales, and his brother Albert, Duke of York, had each been honorary life members of the Club since 1924 and, when it was announced that the Prince of Wales would be visiting Cardiff on February 5th, 1927, Colonel Homfray, as the Club's President, wrote to the Prince's Equerry offering the Prince the facilities of the Club during his stay in the city.

The invitation was accepted and the Prince of Wales joined a host of members for lunch, which was set at a fixed charge of 3/6d. In the absence still of permanent bar facilities, a refreshment bar was set up in the front lounge and the committee room, whilst liqueurs were served from 2.30pm in the main hall. It was a very grand affair which left a sizeable loss in the provisions account, but the following year, when visiting the city on June 12th, the Prince declined an offer to use the Club's facilities once again.

The next royal event was recorded for many years in the entrance hall of the Club by a framed letter, dated May 29th, 1936, from Wigram, Keeper of the Privy Purse at Buckingham Palace. The letter addressed to Colonel Homfray says: 'I am commanded by the King to inform you that His Majesty has been graciously pleased to grant his patronage to the Cardiff and County Club.' The King in question was Edward VIII whose abdication did not take place until December 11th, 1936. As Duke of Windsor, he continued to be Patron of the Club until after the Second World War.

The royal patronage was a great fillip to the Club and its drive to recruit new members. Among the many new members of the Club was stockbroker Freddie Mathias whose offices adjoined the Fire Station and looked down on the Club. Born in Abercynon, Freddie was the grandson of William Henry Mathias, a railway entrepreneur and mining magnate who owned many collieries in the Rhondda Valley. Freddie's father, James Mathias, had

PRIVY PURSE OFFICE,
BUCKINGHAM PALACE, S.W.

29th May 1936.

Dear Sir,

I am commanded by The King to inform you that His Majesty has been graciously pleased to grant his Patronage to the Cardiff & County Club.

Yours truly,

Wigram

Keeper of the Privy Purse.

A letter to the President of the Club from Buckingham Palace.

become the manager of the mining enterprise, as William diversified his interests and, in 1901, he went into a partnership with the Windsor Steam Coal company whose chairman was George Insole, the former President of the Club, whose business interests also included collieries in the Rhondda and whose sporting interests included hunting and cricket. A few years later, the Mathias family also became involved in a milling business in Cardiff with the thriving operation run by Spillers, as well as overseeing the operation of various quarries in the Vale of Glamorgan and becoming a major shareholder in companies supplying electricity to the thriving industrial settlements in south Wales.

Freddie was one of five children born to James Mathias and his wife Eveline James and, after initially boarding at Cowbridge Grammar School, Freddie went to Clifton College where he won a place as a spin bowler in the school's 1st XI. On leaving Clifton in the summer of 1916, Freddie joined the Royal Flying Corps as a Second Lieutenant. He was one of many recruits straight from school as the RFC swelled its ranks and, like his fresh-faced colleagues, he had shown great prowess in flying gliders at Filton Airfield when serving with the Training Corps at the famous public school in Bristol. He passed his flying exams at the Beatty Flying School at Hendon in January 1917 flying a Caudron biplane, but local legend has it that, shortly after qualifying, he crashed the last surviving training plane, much to the glee of other trainees who had found it such a devil to fly!

He was then sent on active service to the Western Front with Flight C of 34th Squadron, and the natural aptitude of the 18-year-old for flying the Sopwith Camels and Sopwith Pups – as much as his fearless attitude – saw him rise quickly to the rank of Temporary Captain. The two-seater bi-planes were equipped with a machine gun, housed above the heads of the airmen and operated by Freddie's navigator, a Welshman called Sylvester. Though many of the new recruits soon lost their lives, the two Welshmen enjoyed a series of successful sorties against the Imperial German Army Air Service which included Manfred von Richthofen, the notorious Red Baron. Though there are no records of Freddie directly engaging fire with the Red Baron, his aircraft may well have been one of several planes which saw action against the famous flyer and his colleagues in the Jagdgeschwader 1 flying unit.

During 1918 Freddie came face-to-face with one of the German pilots after one had to make an emergency landing near the 34th Squadron's base. As the German scrambled out of his machine and dusted himself down after his landing behind Allied lines, Freddie was in a small group of British airmen who had been summoned by the Brigadier on duty to go and capture the German pilot. The Brigadier himself accompanied the party, each man armed and ready to use their guns if the pilot resisted capture. But Freddie and his colleagues could not have been more surprised when the Brigadier, after ordering the German to remove his helmet and goggles, said: "Good God, it's Fritz – we were at Cambridge together!"

As the two men cordially shook hands and exchanged pleasantries, Freddie quietly walked up and removed the German's Luger pistol. The old pals duly walked back chatting away merrily, while the gun remained

Freddie Mathias, as a young pilot in December 1916.

in Freddie's possession for many years, and had several airings at the Club, with Freddie wreathed in smiles and laughter as he recounted the events associated with his disarming of a German airman.

There was nothing humorous, though, about Freddie's other sorties in the ensuing months, including those in September 1918 when, at the age of just 20, he won the Military Cross for his gallantry after completing many hours of successful reconnaissance flights over enemy lines in France and Belgium. He had often flown without any air cover, with a camera having replaced the machine gun in front of Sylvester, his navigator. When taking off and flying towards enemy lines, they had an armed escort, but often when taking photographs over the combat zone the only protection the intrepid pair had was from a rifle which Sylvester had taken on board.

From their base in Northern France, the pair took a superb series of images of the German lines, allowing the Allied forces to know exactly where to attack and, given the presence of dummy soldiers and decoy lines, where not to go. Freddie's citation for the Military Cross paid tribute to his 'conspicuous devotion to duty having carried out several successful shoots which did considerable damage to the enemy. He also successfully took a large number of photographs and obtained much valuable information.' He was subsequently presented with the prized medal by King George V, and joined a list of relatively few airmen who won this award, as distinct from the Distinguished Flying Cross.

Given the fact that life expectancy for pilots on the Western Front was only about eight weeks, Freddie did very well to return home unscathed. Indeed, the only direct attack on his person, as opposed to his plane, had come from a woman who had struck him with an umbrella on the London Underground during a Zeppelin raid. At the time, Freddie was training at

Freddie Mathias (sitting, second left), with fellow members of C Flight of 34th Squadron in 1916, including Sylvester (standing, first left).

Croydon Airfield and was wearing his pilot's uniform, but the lady was in no doubt about what he should have been doing, and after striking him with her umbrella, she yelled "You should be up there young man, fighting the Hun!"

After the war was over, Freddie went up to Cambridge University and read geography at Gonville and Caius College, where he represented the college at rugby, cricket, football and won a place in the rugby club's rowing eight. Although he did well on the cricket fields for the college, he failed to make the Light Blues XI but, during the university vacations, he met with further success playing for Cardiff as well as playing rugby for Glamorgan Wanderers.

It was during one match for the Wanderers, against Clifton RFC, that Freddie travelled back home to south Wales by train sitting on the roof of a carriage. His feat – either daring or foolhardy depending on your outlook – was sufficient for the youngster to gain entry to Cambridge's notorious Narkover Club whose membership required the undertaking of a daft deed. After surviving the spats with the Red Baron and other German pilots, to say nothing of his other sorties over German lines, riding on the roof from Severn Tunnel Junction to Newport must have been a doddle.

During his days as an undergraduate Freddie was often called into the Glamorgan team and, in September 1922 before returning to Cambridge, he joined his county colleagues Johnnie Clay and wicket-keeper Mervyn Hill on the MCC tour to Denmark, which comprised three matches on a matting wicket in Copenhagen. After coming down from Cambridge he became a stockbroker based in Westgate Street – a very convenient location given his sporting interests and membership of the Club – and if Glamorgan ever found themselves short, Freddie was on hand to turn out.

He continued to play cricket for Glamorgan until 1930, and fully enjoyed the camaraderie of life as an amateur on the county circuit, especially the socialising with other amateurs. In the case of the rain-affected match with Lancashire at the Arms Park in May 1927, this led to a bit of a spat with Ted McDonald, the Tasmanian fast bowler who was playing for the Red Rose county. Only a couple of overs in the Glamorgan first innings were possible on the opening day of the contest and, shortly after the umpires had called play off for the day, Freddie and fellow Club member Johnnie Clay took a group of the visiting cricketers out to Cowbridge where they had several drinks in the market town's pleasant hostelries. But, as the evening wore on, the conversation between Freddie and the fast bowler become less and less convivial to the extent that, after Freddie had wagered the Australian that he wouldn't get him out the following morning, the Lancastrian replied amid other fiery oaths that he would knock his block off the next day when play resumed.

Freddie was largely undeterred by the threats and when his turn came to bat the following afternoon, McDonald was true to his word as he unleashed a series of short balls against the Glamorgan number eight, who politely smiled back every time the ball whistled over his shoulder. With steam almost coming out of his ears, McDonald

Freddie Mathias (far right), with Johnnie Clay, Norman Riches and Trevor Arnott during a Glamorgan match against Derbyshire at Chesterfield in 1926.

unleashed another thunderbolt which Freddie hooked for four, with the local journalists writing that 'he almost played the ball off his eyebrows.' The Australian's response went unrecorded but, shortly afterwards, it started to rain again, washing out the rest of the day's play and with rain still falling the following morning, the umpires abandoned the game, although at least one member of the Glamorgan dressing room was keen to have another go at the Lancashire attack!

F for Freddie, F for Fun

Either side of the Second World War, Freddie Mathias was the life and soul of many a party at the Club as well as the instigator of several practical jokes within the Clubhouse. Indeed, it was he who was responsible for releasing a white mouse in the dining room near where Sir Robert Webber was dining.

"Good God, I think I've just seen a mouse," was Sir Robert's response. Much to the glee of Freddie and his dining pals the waiters and waitresses started to scurry around in search of the rodent. After several minutes of frantic searching, Freddie calmly and unobtrusively caught his pet.

He also retained his membership of Cardiff Flying Club and, together with his good friend and fellow Club member G.V. Wynne-Jones, he spent many hours in the air. Freddie also wooed Eileen Davies – the daughter of a coal magnate who lived at Radyr – by flying over the hills and valleys of south Wales, before venturing out over the Severn Estuary.

It was in rather marked contrast to their first flight together, which had seen the pair end up with a forced landing in the grounds of a country house in Northamptonshire owing to fog. Freddie continued to be a regular face at the Club until his death in April 1955 and, in the words of his obituarist, 'F was for Freddie as well as for Fun. He was endowed with whimsical humour, an impish nature and a loveable character.' The Club and many other places in south Wales were much quieter after his passing.

14

Motor Cars and Bodyline

"The Royal Porthcawl Golf Club – it's a bit like the Cardiff and County Club by the sea!"

The hustle and bustle of the main streets of central Cardiff during the 1920s reflected in many ways life before the Great War with shoppers gazing into the shop windows, clerks and other white-collar workers hurrying to their offices, and other traders eager to make their money through a sale. But there was one major change in that motorised vehicles had started to replace the horse-drawn vehicles. Indeed, by the 1920s, many Club members either drove themselves, or had a servant drive them, to the Club in Westgate Street.

Several prominent members had been among the pioneers of motoring in the city before the outbreak of war. On New Year's Day 1904, legislation was passed requiring all vehicles to carry a registration number and 21 vehicles were duly registered in Cardiff, with the first belonging to John Courtis, the Australian-born businessman who owned the brickworks to the north of the city which had supplied the Bute and Tredegar estates, as well as other building contractors, with the material for the wave of house-building that took place during the late 19th century. Courtis joined the Club in 1912, besides serving as a town councillor and later the city's Mayor.

> ### The Club's first driver?
>
> There is no known record of who was the first Club member to drive along Westgate Street and park their vehicle outside the Club, so the honour might have gone to Colonel Homfray, whose Daimler was amongst the first batch of registered vehicles in Glamorgan.
>
> Another early driver to and from the Club was Jack England, a member of the family of potato importers based at West Bute Dock.

His grandfather John Humphrey England had been a member of the Club since its earliest years in the late 1860s, with Jack following suit in 1920. Before the Great War, Jack had gone into Cardiff from his home at Pentre Gwilym in Thornhill by horse-drawn buggy, before becoming one of the motoring pioneers of the area. Indeed, he is reputed to have been the first person to drive a car over Caerphilly Mountain.

Edward England on his way to work from his home at Pentre Gwilym.

During his years as a member of the Club, the trade of Edward England and Company blossomed, with the firm having their own 322-tonne ship, called the SS Cardiff City, built in Paisley, which carried potatoes, fruit and other vegetables from France, Ireland, Belgium, Holland and Poland.

Jack England taking part in motor trials in his Talbot, during the 1930s.

> Robbie Norris, Jack's grandson, is currently the fifth generation of the England family to be associated with the Club. A member since 1975, Robbie runs the Glyndŵr Vineyard in Llanblethian with his brother Richard. What began as a hobby and in Robbie's own words, "a bit of fun", led to the first bottles of wine being produced in 1981. Nowadays, there is an average output of 10,000 bottles, with Waitrose, Sainsbury's, the Wales Millennium Centre and the Welsh Assembly all being customers of the wine from Robbie's vineyard.

Robbie Norris (front right), and his brother Richard at Edward England's 150th anniversary celebrations in 1992.

After the war, the number of drivers arriving at the Club steadily increased and, by the late 1920s, it had reached such a level that in February 1927 the Club's Secretary was instructed to arrange a meeting with the Co-operative Wholesale Society with a view to members parking their cars on the society's land opposite the Club.

Nothing came from this meeting so attention turned to the garden, and other land at the rear of the Club, where vehicles could be parked. Colonel Homfray, in his capacity as President of the Club, contacted the Bute Estate in February 1928 about acquiring a portion of land from the Estate which could be used as a roadway to the garden which in turn could be converted into a garage or car park. The plan did not meet the favour of all members, several of whom had enjoyed the garden parties and other functions which the Club hosted during the summer months where many had mingled with their wives and daughters.

The Bute Estate, however, were still looking to make money from their land in the capital and their solicitors swiftly drafted an agreement, but the asking price was deemed unacceptable and, after much discussion, the idea of converting the garden was shelved. In the meantime, a roadway was created to the south of the Club down to the rugby ground, and members started parking their vehicles there, sometimes to the annoyance of the Athletic Club's ground staff.

On February 27th, 1934 a Special Meeting of the Club was convened at which Hugh Ingledew proposed that the committee should 'carry out a scheme for adapting the Club garden for use as a parking place for members' cars.' Reflecting the unease about such a move, Ivor Downing proposed an amendment that 'such use of the garden would in many ways be detrimental to the Club and its present members.' The Vice-President Sir Harry Cousins could not attend the meeting but he wrote a long letter, opposing the use of the garden and raised the point that those members

The rear of the Club and the garden.

On the links at Southerndown in the early 1900s.

who were chauffeur-driven to the Club, could not practically expect their drivers to stay all day at the rear of the Club.

Other concerns were expressed about potential damage to vehicles when parked on Club premises and who was liable. Others felt that it could be dangerous if cars were driven late at night from the rear of the Club. Downing's amendment was defeated but the strong 'anti' lobby meant that the whole matter was deferred until the AGM on April 28th, 1934. In the meantime, Harry Cousins contacted the Bute Estate and an agreement was reached which preserved the garden at the rear of the Club, with members now being given formal permission to park on the roadway into the Arms Park.

While the inter-war years saw many changes in and around the Club, there were several changes as well to the pastimes and interests of members. Through the actions of Arthur Ingledew, and the generosity of the Glamorganshire Golf Club in Penarth, golf had been a popular recreation for members in the years leading up to the Great War. After the war, John

Duncan was the leading figure among the golfers at the Club besides being the managing editor of the company which published the *South Wales Daily News* and the *South Wales Echo*. In his youth, he had played rugby for Cardiff but golf became his passion, and he won the Welsh Amateur Championships in 1905 and 1909.

Following the creation of the Southerndown Golf Club in 1905, John and his family began a long association with the club. Indeed, this area running west along the coast from Ogmore-by-Sea to Porthcawl became a popular summer retreat for many Cardiffians, and John Duncan took great delight in organising matches between Club members on the Southerndown course.

An excellent golf course also existed in Porthcawl, thanks to the support of another eminent Club member, Edmund David, during the mid-1890s. As agent to the Margam Estate, David oversaw the creation in 1895 of an 18-hole course in the resort town on land owned by the Mansel Talbot family. At the time of its opening, it more than replaced the nine-hole

John Duncan (front right, sitting) with other members of the Glamorganshire Club at Conwy in 1905.

course which a group of coal and shipping magnates, who were members of the Exchange Club, had created on Locks Common on land owned by the Porthcawl Vestry. The new course, overlooking Rest Bay, swiftly became popular and had an added bonus when in 1909 King Edward VII bestowed on it the rare privilege of using the prefix Royal.

By the early 1930s a number of matches among Club members had been successfully staged at the Royal Porthcawl Club and, together with the games at the Southerndown Club, there was sufficient interest to formally establish a Cardiff and County Club Golfing Society. Created on July 24th, 1934, the annual subscription was five shillings, with two meetings a year and matches against other Clubs. Given his early involvement, it was fitting that John Duncan was one of the winners of the meeting held at Royal Porthcawl in November 1935. He went on to become chairman of the Welsh Golfing Union, and president of the Glamorgan County Golf Union, whilst his second son Tony subsequently became one of Wales' leading amateur golfers, captaining the Great Britain and Ireland team in the 1953 Walker Cup.

The annual lunch of the Cardiff and County Club still ends with toasts to the Duncan family, with port poured from a decanter inscribed with the words 'The Duncan Dynasty'.

Two other leading members of the Glamorganshire and Royal Porthcawl golf clubs were Tom Barlow and Joe Simpson. Both were Welsh rugby internationals, and cricketers with the Cardiff club. In the case of Barlow, he also played county cricket for Glamorgan, before becoming a solicitor in Cardiff and a leading figure in the Welsh Golfing Union. The pair

The committee of the Welsh Golfing Union at the Welsh Amateur Championship held at the Glamorganshire course in 1902. Tom Barlow (sitting, far right) next to Joe Simpson, who was chairman of the Union at the time.

were firm friends for many years and considering the amount of time they spent at the Arms Park, it was no surprise that they became members of the Club. Indeed, they would often play at the Porthcawl course before travelling back by train to Cardiff to dine at the Club before heading home.

In April 1901, Joe Simpson invited Ivor Lewis, a well-known JP

Driving off the first tee at the Glamorganshire course in 1889, shortly after the Golf Halt (seen in the background) had been opened on the Barry and Penarth railway line. Tom Barlow is watching the action with his caddy.

and doctor from Porth, to share a round on the Porthcawl links, followed by dinner at the Club with two other friends.

Dr. Lewis' decision to join his sporting pal at the Club would prove to be a tragic one as, after having a few drinks, they dined on mussels soup, followed by steak. However, during the main course, Dr. Lewis was taken ill. What appeared at first to be a bout of indigestion turned more serious and, after vomiting several times in the lavatories, he went to the smoking room for a lie down. A first-aider was summoned from the nearby police station, before messengers were sent out in hansom cabs to fetch Charles Vachell, the town's leading medic, who lived in Charles Street. But, by the time medical help arrived, Dr. Lewis had died, with the inquest finding that he had died from a severe allergic reaction to the mussels. Indeed, Simpson and others testified that Dr. Lewis had never previously eaten mussels and whilst dying in the smoking room, had repeatedly said in a hushed whisper, "I'm poisoned, I'm poisoned."

Another magnate in the newspaper world, Sir Robert Webber, had been one of the most influential figures among the Club's membership advocating the creation of a Golfing Society. As a member of both the Southerndown and Royal Porthcawl clubs, as well as those at Radyr, Walton Heath and

David Webber who, aged nine, was en route to Canada when the vessel he was travelling in was torpedoed by a German U-boat. He spent two days in a lifeboat before being rescued.

the Royal and Ancient at St. Andrew's in Scotland, he became a leading figure in its affairs, donating in 1935 a splendid silver cup, known subsequently as the Webber Cup for the annual competition held by the Golfing Society in June, with Sir Robert's trophy being presented to the winners.

After entering the newspaper world during the early 20th century, Sir Robert lived in Cyncoed for many years with his wife Jane and daughter Joan. In the words of one of his closest friends, Sir Robert was 'a bundle of vim and vigour' and as a young man he was the driving force in 1912 along with Reuben Pugsley and Charles Hardwicke in the creation of Cardiff Business Club, with Sir Robert serving as its first President.

As managing director of the Western Mail and Echo Ltd, he was also a man of some influence and, in 1922, persuaded the Great Western Railway to amend their late evening timetables to fit in with the printing schedule of the newspaper. These alterations allowed, for the first-ever time, copies of the *Western Mail* to be on the breakfast tables in north Wales, and as its sales rose, the *Liverpool Post* was forced to drop its price from 2d to 1d in a bid to win back trade. Another measure of his management skills came during the General Strike of 1926 when the *Western Mail* and *Evening Express* both managed to bring out editions on each day of the strike. Sir Robert's efforts in maintaining this service were later commended by Prime Minister Stanley Baldwin.

Sir Robert was knighted in 1934 for services to the newspaper industry and to south Wales, and besides a special dinner with his close friends in the Club, 1934 also saw him honoured with a civic dinner in The Park Hotel. He later became chairman of the Cardiff Bench and, as Deputy Lieutenant, Robert was very well-connected in many aspects of life in the City of Cardiff and, on the retirement of Sir Rhys Rhys-Williams in 1950, he became the Club's President. He held this office until his death in 1962. Even in his later years, Sir Robert

Sir Robert Webber.

took an active interest in the golfing affairs of the Club besides dining on a regular basis with his friends, usually at the far window table. Indeed, it was here where he was often heard to say to his guests 'the more you eat and the more you drink, the more we like you'.

During the inter-war era, Sir Robert was also the proud owner of a Humber Snipe which bore the registration plate ANY 1 and, after he died, the plate was transferred to Jeffrey Archer, the author and politician. Sir Robert was also one of the members to arrive regularly at the Club – or to one of his favourite golf courses – by chauffeur-driven car, with either Andrews or Harding taking Sir Robert from his home in Cyncoed into Cardiff. The Humber Snipe, and its driver, parked at the rear of the Club while its owner wined and dined with his pals in the Club.

During this time, there were other features created next to the car park including a new squash court. The closure of the adjoining Racquets and Fives Club after the First World War, as well as a growing interest in squash, had prompted the committee on April 29th, 1933, to agree to the installation of a portable squash court in the garden at the rear of the Club. The prime mover behind the scheme was Captain W. R. Bailey and, after consulting the Berkshire Club, who had put up a similar court at their base, £300 was spent on the completion of the structure. The fee for using the court was one shilling for 30 minutes, followed by a bath in the washroom on the upper floor of the Club.

Judging by the accounts for 1934, the new court was a popular addition to the facilities with takings totalling £40 and five shillings. The expenses of running the court amounted to a fraction over £22, but the profit made on the facility was soon wiped away as the court began to take in rain water. Captain Bailey approached the committee in December 1938, asking for £55 to undertake a couple of structural alterations plus the addition of a waterproof seal, but his request was rejected and the portable court was sold in January 1939 for £50.

Hunting and horse racing remained popular interests of Club members during the inter-war period, and the race days at Cardiff led to some lively conversations in the evenings and days after the races. The Club's Trustee, Bob (R.H.) Williams of Bonvilston House continued to be a prominent figure in hunting and steeplechasing circles, having been an energetic Master of the Glamorganshire Hunt since 1914 and had carefully guided the Hunt through many difficult years during the 1920s before standing down in 1934.

His son Lewis Williams also became a well-known face at the Club after leaving the Oratory School and starting to work in Cardiff. Known to everyone as 'Tip', he was a great character and talented amateur sportsman.

'Tip' Williams.

Through his friendship with fellow Club member Johnnie Clay, he played several times for Glamorgan when, as a result of financial difficulties, they were looking to augment their side with decent amateurs, rather than spending what little money they had on the services of professionals.

'Tip' Williams was also one of the founder members of the South Wales Hunts Cricket Club in 1926 and he revelled in the more social form of the game, playing alongside several of the Cardiff and County Club members. He also enjoyed chatting about racing with his great friend Johnnie, as well as attending the meetings at Chepstow Racecourse where Johnnie took great delight in entertaining his many friends from the world of sport, business and the Club. Another good friend was Maurice Turnbull, the son of the Cardiff shipping magnate and the man who, from 1930, took over the captaincy of Glamorgan CCC. Educated at Downside School and Trinity College, Cambridge, Maurice made his debut for the Welsh county in 1924 while still a schoolboy and, during the winter of 1929-30, he became Glamorgan's first Test cricketer as he made his England debut in the Test Match against New Zealand.

Maurice had been a member of the Club for several years and sometimes, after play at the Arms Park, he would invite the visiting captain to dine with him at the Club. On occasions, he and Johnnie Clay invited other players into the Club as their guests, but their decision in May 1936 to welcome some of the visiting Yorkshire side led to some caustic complaints, with the Management Committee expressing concern that professional sportsmen had come into the Club.

Maurice also dined after play in The Grand Hotel. It was here in August 1932 that he and the Nottinghamshire captain ate after a remarkable day's play which had seen the visiting bowlers, Harold Larwood and Bill Voce, experiment with fast leg-theory bowling ahead of the MCC winter tour to Australia. What subsequently became known as 'Bodyline Bowling' was treated with great disdain by Turnbull, and his batting partner Dai Davies, who time and again cut and pulled the ball for four, much to the delight of a large crowd who had thronged into the cricket ground in anticipation of seeing the Nottinghamshire bowlers' new tactics. Maurice completed a double hundred, and Davies posted a century, as Glamorgan passed the 500-mark for the first time in their history.

While Maurice and his counterpart were enjoying a fine meal in the Grand Hotel, the Nottinghamshire bowlers were drowning their sorrows in neighbouring watering holes. After fully quenching their thirst, they decided to return to the cricket ground and to express their feelings about what they perceived to be a 'feather-bed' surface. Despite being followed by a couple of reporters, the visitors showed their contempt by urinating on the wicket, much to the anger of the Arms Park groundsman who sent a message to the staff in the Grand Hotel about the nocturnal watering.

It was late when Maurice went back across to the Arms Park and, despite being exhausted after his batting exploits, he was upset at the actions of the visiting bowlers. He was also worried that the *Western Mail*'s correspondent had got the story about the night-time watering and the roughing up of the groundsman who had tried to stop the prank. He therefore walked across to Park Street and the *Western Mail* offices, where his contacts with Sir Robert Webber came to good use, as the story about the nocturnal visitation was removed from the morning editions of the newspaper. A posse of photographers were present the following morning when the covers were removed at the Arms Park, but the ground staff had worked their magic in removing the rust-coloured patches of turf and the story remained a secret.

As well as being an excellent diplomat, Maurice was a highly talented all-round sportsman. Besides playing cricket for Glamorgan, he was a gifted scrum-half for Cardiff RFC, and won a couple of Welsh caps, including the match against England in February 1933 when Wales won for the first time at Twickenham. He also played hockey at a high level, was an active member of the Club's Golfing Society and was a founding member of Cardiff Squash Club. His sporting activities, his work in an insurance brokerage with his brother-in-law Ted Glover, and regular articles for the *Welsh Catholic Times* all gave Maurice a wide circle of friends from every aspect of life in Cardiff.

Maurice was therefore not short of people to invite into the Club when Wales played their home rugby

Maurice Turnbull (right) in his office, with Ted Glover.

internationals at the Arms Park. For several years, the practice had been for every member to have two guests for admission to the Club by ticket only. Guests were entitled to use any part of the Club's premises but ladies were only allowed on the roof balcony. A dozen tickets were also sent on each rugby international day to the Lord Mayor for his use and those of his guests. The Club's stewards on duty on match day, as well as other staff, were instructed to keep a close eye on the wooden door at the bottom of the garden at the rear of the Club which led directly into the rugby ground. There were strict instructions that while the Lord Mayor and his guests were very welcome to watch the game from the balcony, if they subsequently went through the door into the rugby ground it was to be locked immediately afterwards and not opened again for their return.

Besides ensuring that the Club's facilities were not abused, even by those in high rank, the Managing Committee also made regular contact with the Athletic Club, and subsequently the Welsh Rugby Union, to ensure that the view from the rear of the Clubhouse was not disturbed. The view over the rugby ground was one of the perks of Club membership, and from its inception in 1922, the Cardiff Athletic Club had received an *ex gratia*

An aerial view of the Arms Park and the Clubhouse during the inter-war period.

contribution to its funds from the Club largely to preserve this unrivalled view. Indeed, in 1927 the Athletic Club were informed by the Club that the annual payment had only been made provided that the stand, recently created by the Athletic Club for people to view the greyhound racing, was not raised any further and that no additional obstructions were put in place.

Details of the unrivalled view from the veranda of the Club had also reached the ears of executives from Screen Productions Limited who in 1929 approached the Club for permission to take pictures of the action from the balcony. It didn't take long for the Management Committee to refuse this request!

15

The Alexanders

"A highly regarded meeting point for Cardiff's business and professional community; a crossroads for a wide group of people of experience and influence in almost every sphere."

Many families from Cardiff and the Vale have a strong affiliation with the Cardiff and County Club so it could be invidious to devote a single chapter to just one. But the lives and loves of five generations of the Alexanders of Gileston near St. Athan, highlight the special place that the Club has played in the history of Cardiff's leading firm of auctioneers and chartered surveyors.

The Alexander family hailed from Newmarket in Suffolk, before Robert senior moved during the 18th century to become stable manager at Fonmon Castle, the home of the Jones family, south of Cowbridge. Robert senior married Catherine Howells of Gileston where, in the course of the next 18 years, they had 13 children, two of whom died in infancy. Robert junior, their fifth son, subsequently became a grocer in the village of Penmark, with his eldest son John taking over the business in the mid-19th century. Another son, Thomas, helped to run the 770-acre Monknash Farm belonging to Edward Perkins.

The family's grocery business thrived, allowing John Alexander to acquire The Mount in Penmark and to send his eldest son David to Cowbridge Grammar School. Born in 1841, it was David Alexander who began the family's long association with the Club, assisted by their close links with the Jones family, and in particular Oliver Henry Jones, the Oxford-educated barrister, whose father and brother were founder members of the Club. David was also an archetypal country sporting gentleman, enjoying many days of hunting during the winter and then during the summer playing cricket for Cowbridge and Glamorganshire.

After completing his training as a surveyor, David Alexander moved to Pontypridd where he set up an auctioneer's business, possibly having been inspired as a young schoolboy by seeing the wielders of the gavel in action at the weekly markets in Cowbridge. It was a shrewd move as his business thrived and he soon became friends with William Prichard Stephenson, a Yorkshire-born auctioneer who also had a successful auctioneering business based in Cardiff, plus a branch in Pontypridd. After the death of Stephenson's first business partner, David Alexander agreed in 1877 to move to Cardiff, and Stephenson and Alexander of 5/6 High Street came into being.

The business went from strength to strength, with David's staunch support of the Liberal cause helping to secure, for his practice, a role as advisers to the town council about the value of land and its suitability for various purposes. His training as a surveyor proved to be a great asset to the business, and complemented Stephenson's strengths which lay with the paperwork, having trained as a clerk with his grandfather Joseph Davies who was a house agent based in the prosperous suburb of Crockherbtown on the eastern edge of the town.

David Alexander as depicted in a caricature in the Western Mail by J.M. Staniforth.

David drew on his background in surveying to advise the Liberal-dominated council as they pondered the various uses of land, potentially for house-building, new industrial premises or for open space and recreation grounds. The 1875 Cardiff Improvement Act had given the town's corporation the power to provide public pleasure grounds, so it was a good time for the go-ahead and industrious surveyor and auctioneer to be working and living in Cardiff. He was able to advise the council on their negotiations with the Marquess of Bute and Lord Tredegar, as well as with the Shirley and Mackintosh Families about developments in the Roath and Cathays area. His social connections with the families and their land agents via the Club no doubt assisted negotiations.

Another pal from the Club, whom David worked closely with, was Charles Thompson, a wealthy corn merchant who was a director of Spillers and lived at Penhill House, adjacent to the junction of Pencisely Road and Llandaff Road. As the suburb of Canton rapidly developed, Thompson became concerned about the lack of space for recreation and leisure. After advice from David, he generously let the locals, from April 1891, use the

Hubert Alexander.

southern part of his land adjacent to Romilly Road. Four years later, the pair worked on a project whereby the northern part of Thompson's land was formally laid out by the landscape gardener William Goldring, with an area of walks and woodland. In 1911 and 1912 David again advised Thompson as the area was handed over to the city council as a public park.

By this time David had overseen the construction of Bryneithin House in Dinas Powys and it was here in 1905 that he and his wife threw a grand party on the spacious lawns of their home to celebrate his appointment as national president of the Chartered Auctioneers – a measure of his influence and standing. By this time, his sons Hubert and John had joined the family's practice, as well as the Club, and were starting to make a name for themselves in and around Cardiff.

Educated at Tavistock Grammar School and Sherborne, Hubert was also a talented all-round sportsman, playing rugby for Newport, Penarth and the Barbarians, and cricket for, among others, Cardiff, Penarth, Dinas Powys and Glamorgan. He would regularly stroll across from the High Street offices on Wednesday afternoons to the Arms Park to watch the rugby or cricket, before unwinding with his friends in the Club. Indeed, in 1886 he had been among the crowd at the Arms Park to cheer on his nephew, Edward Perkins Alexander, when he played for Wales in the Home Nations Championship against Scotland. The former pupil of Llandovery College and Sherborne School had first been chosen as a forward in the Welsh side when a student at Jesus College, Cambridge, and Hubert took great delight in watching his nephew wear the red jersey.

Hubert married Edith Duncan, the daughter of newspaper magnate John Duncan, with the pair having first met at a Club dance. They subsequently lived at Gileston Manor in the Vale of Glamorgan, with Hubert adding golf to his recreational interests and taking part in the Club's golf days at Southerndown. Hubert became a partner in the practice shortly before the Great War and he duly followed in his father's footsteps by serving as national president of the Chartered Auctioneers. Even when in his seventies, Hubert would drive daily into the Cardiff office from Gileston Manor before often having lunch at the Club. But, on the morning five days before Christmas in 1954, he turned around and headed back home as he felt unwell. Sadly, he died later that day.

The Alexanders were in mourning again the following year as John died in 1955. As the older brother by a couple of years, he had become the senior partner in the firm in 1922. He also played cricket for Cardiff but, unlike Hubert, he remained a bachelor and spent his time during the week living in rooms at The Angel Hotel and socialising at the Club, before spending his weekends at West Cottage, Southerndown, where he took great delight in having a round of golf on the course run by his sister-in-law's family.

John was a larger-than-life character and despite living literally just around the corner from the offices in High Street, he would always arrive by car each morning. Indeed, it would be the job each weekday of one of the five trainee clerks in the office to walk to the Queen's Garage, opposite the Clubhouse in Westgate Street, and drive John's car up to The Angel Hotel where at 9.30 precisely each morning he would walk down the steps of the hotel and stand under the portico, before getting into the Daimler. His daily ritual brought a wry smile to the faces of his chums in the Clubhouse. But as John would frequently say, it taught the trainees to be punctual and prompt, and after an ear-bashing from the garrulous auctioneer, they would never be late a second time!

Hubert Alexander (right) with Colonel Henry Lewis of Greenmeadow at the Cowbridge Show.

Routine was clearly an important aspect of John's life especially with the journey to his weekend retreat. Every Friday teatime, John travelled by train from Cardiff General to the Southerndown Road station where he was met by Wilf Powell, his housekeeper. Before the Second World War, Wilf met John in a pony and trap, but after hostilities were over, he drove a Bentley to and from West Cottage. When it came to dining in the Clubhouse, John would always sit at a table by the door leading into the dining room, sitting on a chair with his back to the door, from which he would meet and greet everyone who entered or left. A few jovial words were always passed on to his fellow diners, and woe betide anyone who tried to sit in what became known as 'John Alexander's Chair.'

After his death, Duncan Alexander, Hubert's eldest son, became the senior partner and the next member of the Alexander clan to play a prominent role in the life of the Club. Also educated at Sherborne, as well as at Trinity

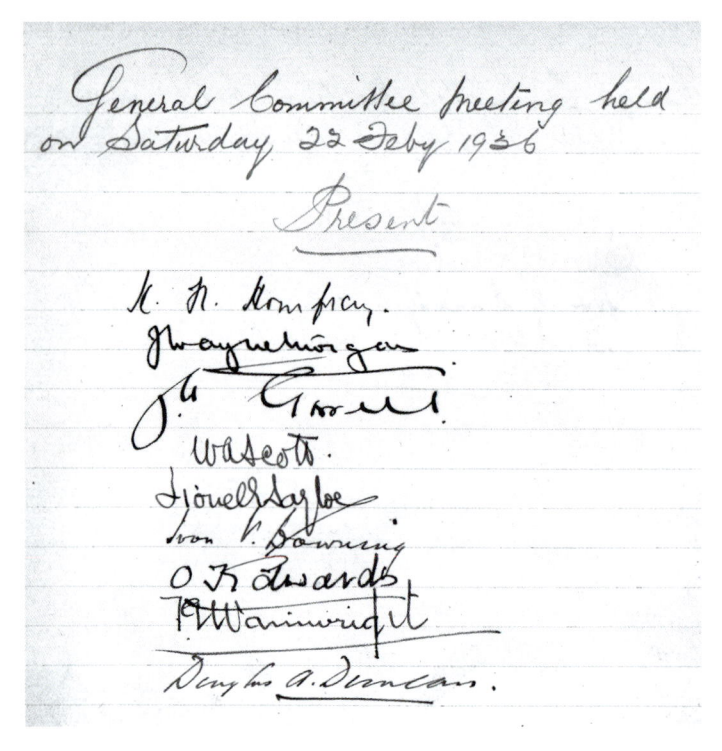

The register for a committee meeting in February 1936.

Duncan Alexander.

College, Cambridge, Duncan continued the family's good name on the sports field, playing for both the 1st XI and 1st XV at both school and college, besides following in his father's boots by playing on the wing for Newport RFC between 1932 and 1936, as well as for Penarth RFC.

In 1939 he married Bobby Hann, the daughter of coal magnate Edmund Hann, the director of the Powell Duffryn Company based in Aberdare, who was also a member of the Club. In 1923 Hann moved from the Rhondda to live at The Rise in Llanishen, with Bobby first meeting Duncan at a golf tournament organised by the Club. The couple lived initially at Tai-on House in St. Andrew's Major, before moving to Star House, Capel Llanilltern, from which Duncan also served as High Sheriff of Glamorgan in 1960.

Despite his heavy workload as senior partner of the company's practice, and his duties as High Sheriff, Duncan played a full role in the social life of the Club and, together with a gang of like-minded pals, would visit Twickenham every other year for the Home Nations Rugby international between England and Wales. Their visit by train in January or February became something of a tradition and led to a host of jolly japes. On one occasion, one of the party had an unfortunate accident in the gentlemen's lavatory as the express made its way between Swindon and Reading. With a rather tight water tap, he applied too much pressure in trying to loosen the valve and rather than washing his hands, a spray of water covered his trousers.

Thinking that Duncan and the others would tease him about the state of his attire, he decided to stay in the lavatory and to dry the wet patch by removing his trousers and dangling them out of the window. For a while all went to plan but, on the outskirts of Reading, he suddenly felt a violent tug from outside the carriage as his trousers caught in the lineside device which grabbed mailbags from Royal Mail trains. The upshot was that the trousers swiftly disappeared out of the window, leaving him standing in his underpants in the cubicle.

Unsure of what to do, he sat on the toilet and waited for his friends to come to the rescue. Concerned at their friend's disappearance for some time, they made contact and hatched a plan to solve his embarrassing predicament. On arriving at Paddington, Duncan and his five pals formed a tight ring around their friend and walked up the platform with their huddle

---

28

BYE-LAWS.

1.—No Member shall take or permit to be taken from the Club, on any pretence whatever, or injure or destroy by writing or otherwise, any newspaper, periodical, book, pamphlet, or other article belonging to or used by the Club. No paper or placard, written or printed, shall be placed in any part of the Club premises without the sanction of the Committee.

2.—No Member shall make use of the address of the Club in any advertisement, and in no business prospectus in any form issued by a Member of the Club shall the name or address of the Club be used.

3.—The following scale of charge for late money shall be paid by Members using the Club after 12 at midnight :—

    1s. from 12 to 1 a.m.
    5s. from 1 to 2 a.m.
    10s. from 2 to 3 a.m.,

the above charges to be cumulative, and the Club to be absolutely closed and lights put out at 3 a.m., and no Member shall be admitted to the Club after 12 at midnight.

4.—No boys under 12 years of age shall be brought into the Club.

5.—No Member or Visitor shall be allowed to bring a dog inside the outer doors of the Club premises.

6.—No umbrella, sticks, or coats shall be deposited anywhere but in the places provided for that purpose.

The Club's bye-laws from 1936.

preventing anyone from seeing their trouser-less friend. From the platform concourse, they deftly manoeuvred their friend into the lobby of an adjoining hotel where they negotiated the purchase of a replacement pair of trousers, allowing the party to continue their journey on to Twickenham. One can only wonder what the British Railways staff thought when they found a pair of gentleman's trousers on the mailbag catcher!

Duncan Alexander also played a leading role in the establishment of a get-together of members of the Hawks Club each year in the Club. The function, which is the only provincial meeting of members apart from the annual bash at The Savoy in London, was the brainchild of Duncan, Clifford Evans and Hugh Thomas, a prop forward for the Light Blues who was taking a postgraduate course in Industrial Sociology at Christ College, and a man who during the 1990s subsequently acted as Chairman of the Club.

Born in Briton Ferry in March 1937, Hugh read history at Swansea University before spending two very happy years at Cambridge. He subsequently worked in personnel management for Metal Box and the Steel Company of Wales, before being employed by Andersen Consulting. His sporting c.v. was equally impressive, playing rugby for Aberavon, London Welsh, Gosforth, Llanelli, Welsh Academicals and Crawshays, and for the past 30 years, he has been able to meet up with many of his friends from the Hawks Club in the annual dinner held in the Clubhouse every January. As he freely admits: "Duncan, Clifford and myself wanted to maintain the

H. Hugh Thomas (front right) after a lunch in the Club with fellow Hawks (left to right), Tony Lewis, Dennis Gethin, John Charles Rees and E. Russell Jenkins.

camaraderie and friendships we forged during our student days, besides finding a way of saying thank you to the University and the Hawks Club."

Duncan shared duties at Stephenson and Alexander with his younger brother Peter, who also attended Sherborne and Trinity, Cambridge. Rather than excelling at rugby and cricket, Peter was a decent athlete and put his prowess at the long jump to good use when serving with the 81st Field Regiment shortly after D-Day. Peter and Duncan had both signed up for military service in the drill hall at Bridgend in August 1939, shortly after hearing that war had been declared. Duncan subsequently worked in the Intelligence Corps in Ireland, whilst Peter joined the Territorial Artillery and trained as a gunner.

When Operation Overlord took place in June 1944, Peter was appointed as one of the gunners on the landing craft heading to Sword Beach on D-Day +10. As they were offloading guns and other equipment for the advance into Normandy, the vessel got lighter and lighter, causing it to drift closer to the shore and away from the landing jetty which others had established. It duly ended up so close to the shore that the landing ramp dropped into the water at least ten feet from dry land. This caused a certain amount of consternation for the officers on board the craft about how to disembark, but Peter swiftly realised that his long-jumping skills would come in useful. After walking up and saluting the commander of the craft, he then ran towards the ramp and safely leapt, *à la* Lynn Davies, on to the beach, allowing him for years afterwards in the Club to regale his pals that he never got his feet wet during the Normandy invasion!

In 1953 Peter married Octavia Verdon Roe, the daughter of the pioneering English pilot and aircraft manufacturer Sir Alliott Verdon Roe. In June 1908 at Brooklands motor racing circuit, Roe had been the first man to take off in a plane with a British engine and his prowess subsequently led to creation of the AVRO company which, by 1918, had three factories across the country, manufacturing thousands of aircraft for the Royal Flying Corps. He was knighted for his efforts in 1929 and, from his home at Hamble in Hampshire, he subsequently showed an interest in the activities of Oswald Mosley and became a member of the British Union of Fascists.

But the actions of the blackshirts were far from Peter's mind when he met Octavia at a 21st birthday party at Oxford. After a courtship of several years

Anthony Alexander.

the pair married and moved into Tai-An House in St. Andrew's Major which had been vacated by his brother, before moving to live at Gileston Manor in 1974. During the post-war years, both he and Duncan were shrewd advisers to the Club as their committee reviewed and assessed the condition of the Clubhouse, besides assessing the viability of other potential properties to occupy should the premises in Westgate Street become unsuitable.

Peter's son Anthony subsequently became the fourth generation of the family to qualify as a chartered surveyor before joining the family firm in 1982, and subsequently the Club in 1986. Anthony attended Highfield Prep School in Liphook before going to Canford School, and then qualifying as a surveyor at the Polytechnic of Wales at Trefforest. After qualifying, he worked briefly for Cluttons in London before returning to Cardiff and looking after lettings of properties, overseen by Stephenson and Alexander, in the various city centre arcades.

In 1987 he married Catherine Pearce, their wedding taking place at Gileston Manor, from which they left by helicopter hired from Julian Verity's company based at the Tremorfa Heliport on the eastern fringe of Cardiff Docks. However, the aircraft was only really a decoy following advice from Robbie Norris, his good friend at the Club, with the Veritair pilot taking them only a short distance to Rhoose Airport where Anthony had left a car. The newlyweds then spent a couple of nights at the Cwrt Bleddyn Hotel near Caerleon before heading to Heathrow and flying to Venice where they spent their honeymoon, and then returned to the UK aboard the Orient Express.

Anthony subsequently left the family firm in 1995 to set up Alexander's Cookshop at the Wyevale Garden Centre near Thornbury. The venture took off and allowed him to sell the business in 2006 and take early retirement. Together with Catherine, he has lived in the Medoc region of Bordeaux in Southern France since 2009, but he retains an avid interest in the affairs of the Club and the city of Cardiff with which his family has been so closely connected for almost 150 years.

16

Austerity and Air Raids

"The Club was a haven of normality and civility at times of war."

In March 1934 the committee of the Cardiff and County Club authorised the purchase of a wireless set and, for the next few years, Club members were able to enjoy, while sitting in one of the upstairs lounges, some of the programmes transmitted by the British Broadcasting Corporation. There was, though, an above-average number of members clustered around the radio at 11.15am on the morning of September 3rd, 1939 when Prime Minister Neville Chamberlain broadcast to the nation and uttered the following words:

> "This morning the British Ambassador in Berlin handed the German Government a final note stating that, unless we heard from them by 11 o'clock that they were prepared at once to withdraw their troops from Poland, a state of war would exist between us. I have to tell you now that no such undertaking has been received, and that consequently this country is at war with Germany."

As in 1914, the news that the country was at war again with Germany came as little surprise to the many territorials and reservists among the Club's membership. For many months, the relationship between Britain and Germany had deteriorated and the entry of Nazi troops into Poland signalled the start of six more long and bloody years of conflict.

By the time of Mr. Chamberlain's historic announcement, a number of Club members were already serving with their battalions and other military units and, over the following months, the footfall through the front door of the Club dropped as the members, and their guests, went off once again to do their bit for King and Country. Maurice Turnbull was just one of several committee members to stand down from their posts with the Club as, in Maurice's case, he swapped cricket whites for the khaki uniform of the

Welsh Guards and, after scoring a century in his final game for Glamorgan away to Leicestershire, he joined hundreds of other brave men as they prepared for battle on foreign fields.

Down at Cardiff docks, anti-aircraft installations had already been created and, though some loose tongues were wagging about a German invasion, plans were drawn up for laying mines, and searchlights were placed on the roofs of some of the tall buildings in both the dockland and commercial areas of the city. The Arms Park once again became a military training ground and, next door in Jackson Hall, the Council's Juvenile Employment Bureau moved out as the building was taken over by the military authorities and earmarked as a base for military personnel if a state of national emergency was declared following a German invasion. Neither took place and for most of the war the building was used by the Admiralty to store telephone and teleprinter equipment.

Within a few weeks of war being declared, petrol was rationed and some Club members, who had previously driven to the Club or been chauffeured by their drivers, complained about the shortage of petrol. In early January 1940, bacon, butter and sugar was rationed, followed soon afterwards by meat, tea, jam, biscuits, cheese, eggs, milk, plus canned and dried fruit.

The food rationing put great pressure on the management and kitchen staff as they tried to keep a semblance of normality in the Club's affairs. By 1940 the amount of meat per week for the Club and its staff was restricted to £2 11s. The response was smaller portions and, even in the years immediately after the end of the war when rationing continued, there were complaints from members about the tiny pieces of meat which were being served. With fuel also in short supply, the Manageress, Mrs. Price, decided to hold back some hot dishes until 1.30pm on Mondays, Wednesdays and Fridays, and the amount of vegetables served was also restricted. However, if members wanted a second helping, they could have it free of charge. Saturday evening meals were also suspended, and evening meals after 8pm were stopped. As rationing continued, evening meals were further trimmed down to just two nights a week, although afternoon tea continued to be served – with sixpence securing a pot of tea, some bread and butter, cake and sandwiches from 3.45pm each weekday.

Drink was also in increasingly short supply, with suppliers' prices also being raised to reflect the shortages. In April 1942 the Club had to start limiting the amount of alcohol served at both lunchtime and on the evenings when hot food was served. Initially, sherry was limited to one glass per member, and there was a limit of two glasses of luncheon port per meal, with vintage port restricted to one bottle between three members.

Not everyone was happy with these restrictions with Ivor Downing, the Club's Vice-President during the war years, and a man who liked to quaff Taylor's 1927, is reputed to have said on being told that port would be restricted to one glass per meal "But I always have three!" In light of these grumbles, a coupon system was introduced, with each member being given six coupons a day – three were issued before 3pm and three in the evening – with each coupon being the equivalent of a single drink.

Despite the fact that many members were away on military service, there was no shortage of people looking to dine or have a drink with friends, as well as others seeking a few home comforts and a place where, for a few hours at least, they could forget about the horrors which were unfolding elsewhere. Under the rules, any gentleman holding a commission in any of the regular forces and being posted within the district could be elected as a temporary member, and as a result there were many new faces in uniform dining in the Club on a regular basis.

Major-General Arthur Solly-Flood.

With supplies becoming more difficult and expensive to acquire, this put even more pressure on the Club and its staff. In May 1940, it was agreed that the subscription for service members away from the district would remain at a guinea a year but, if they returned and used the Club for more than three months, they would be asked to pay a proportionate subscription. Maurice Turnbull, then a Captain in the 1st Welsh Guards, was one of the members to dine in the Club while returning to see family and friends during their leave from military service. Others were Lincoln Hallinan, then a Lieutenant in the Rifle Brigade, as well as Colonel Geoffrey Gaskell and Major General Arthur Solly-Flood.

Pilots and Dambusters

Many of the Club's members were enthusiastic members of Cardiff Aero Club, created at Pengam Moors after the Great War. Among the intrepid and pioneering aviators were Club members Edward England and Charles Keen, whose family bought the East Moors works in 1899 before merging it with other steel-making companies to form the Guest, Keen and Nettlefold company.

As the photograph below shows, both Edward and Charles were delighted in 1937 when Amy Johnson, the pioneering lady aviator, visited the Aero Club, seven years after she had become the first woman to fly solo from Britain to Australia. Sadly, she died in January 1941 while transferring an aeroplane from Prestwick to an RAF base near Oxford.

Amy Johnson (sitting, second from the right) visits Cardiff Flying Club in 1937. Also pictured are Jack England (standing, left) and Charles Keen (standing, right).

Wing Commander Guy Gibson was another famous pilot and visitor to the Club, besides being a frequent dining companion with Freddie Mathias and his friends. The man, who later became the leader of the legendary Dambusters raid over Germany in May 1943 and won the Victoria Cross for his efforts, had met a young Penarth girl, Eve Moore, at a party in Coventry during December 1939 while he was on three days leave at his brother's house.

In 1940 the pair got married at All Saints Church in Penarth, with Gibson flying his Blenheim bomber from his airbase in Lincolnshire to the RAF's newly-installed landing strip at Pengam Moors. As a leading figure with Cardiff Aero Club, Freddie had overseen some of the arrangements for Gibson's visit and the pair retained their friendship.

Austerity and Air Raids

> They had a common love of flying and playing golf. Indeed, Gibson had become an honorary member of the Glamorganshire Golf Club and after the Dambusters raid, he spent his two weeks' leave in south Wales, playing golf on most days.
>
> Indeed, it was while he was in Penarth that he had a call from the Air Ministry telling him that he had been awarded the Victoria Cross and, on his next leave to south Wales, Freddie had great delight and pride in dining with Gibson in the Club's dining room.

Two years into the war and Cardiff was starting to show the effects of hostilities, following a series of German air-raids over the city. In January 1941, a magnetic mine was dropped on Whitchurch and subsequently the clockwork mechanism was presented to the Club, as a rather unusual trophy of war by Captain Dispard. The following month an air raid on

A cartoon from the Western Mail after a wartime air-raid over the Arms Park. The main stand of the rugby stadium suffered bomb damage, whilst the windows and exterior of the nearby Cardiff and County Club were also damaged.

Cardiff saw several of the windows of the Club being blown out, with the committee deciding against replacing the glass which in any case was in short supply. Instead, the frames were boarded up, made airtight and treated with preservative. The same happened to some of the windows in the dining room and reading room and, soon after, two fire-watchers were employed at the princely sum of £2 per month.

In May 1941 the furniture and other movable objects were insured against war damage, and the Club's insurance policy for fire was increased to £7,000. As air raids continued, the committee also decided that a place of safety should be found for the wine stock. By 1942, the death toll of members and their offspring had also risen, while other Club members such as Wilf Wooller, the Welsh rugby international and Cambridge Blue, were reported as missing in action in the Far East and, with no news for many months, the worst was feared.

Another member held as a prisoner of war by the Germans was Horace Cyril Phillips. 'Phil', as he became known to Club members after the war, was captured in Arras in 1940 and spent almost five years in Stalag 383. After being liberated, he continued his military service by becoming Academy Sergeant Major at RMA Sandhurst before being appointed Sergeant Major of the Queen's Bodyguard, Yeomen of the Guard, and travelling with Her Majesty on a number of overseas tours, as well as on domestic engagements.

In his youth, 'Phil' had played rugby for Newport and London Welsh and, in later life, he became a keen and wily fisherman. Once, upon hearing some bailiff's claim that nobody could ever poach fish on the Wye under their jurisdiction, 'Phil' went out and caught three trout by tickling them, before laying them on the senior bailiff's doorstep with a covering note containing his compliments.

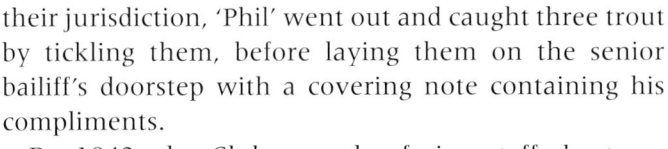

By 1942, the Club was also facing staff shortages and, in April that year, it decided to close the billiard and card rooms, whilst opening hours on Sunday were greatly restricted. Holidays became impossible for the staff and, in return for their long hours, they were paid an extra week's wages for their additional shifts. As the war progressed, staff weariness became more and more of an issue and, in 1943, 1944 and 1945, the Club closed for a fortnight during August so that the exhausted staff could finally have a break.

Nevertheless, the rising pressures and a few spats with the committee saw Captain Price being told that he was going to be demoted to Steward and that another

Colonel Sir Rhys Rhys-Williams.

Club Secretary would be appointed in his place. On hearing this news, he promptly resigned, although his wife briefly continued as Manageress and the general overseer of catering operations before being replaced by a Mrs. Roderick. Arthur Ellis, a long-standing Club member, stepped into the breach and acted as Honorary Secretary from May 1942, and worked alongside Mrs. Roderick for the remaining years of the war.

Much of the day-to-day running and decision-making at the Club was undertaken by Arthur Ellis in partnership with Vice-President Ivor Downing. The President, Lieutenant-Colonel Sir Rhys Rhys-Williams, spent a great deal of time in London serving in the War Office and, in his absence, it was Ivor Downing who was left to deal with the more mundane matters in the Club. Sir Rhys Rhys-Williams had succeeded Colonel Homfray as President in 1940 when he resigned on health grounds on April 4th, before sadly passing away soon afterwards. Rhys-Williams was the son of Judge Gwilym Williams of Miskin Manor and during the Great War had been awarded the Distinguished Service Order in 1915, besides being wounded twice, while serving with the Welsh Guards. The Old Etonian also acted as Assistant Military Attaché in Tehran and ran an intelligence service for the Russians in their campaigns against the Turks and was awarded the Order of St. Vladimir with the Swords by the Tsar in 1916. He subsequently became an Assistant Director in the War Office, with specific responsibilities for general movements and railways.

Rhys-Williams had read law at Oriel College, Oxford, being called to the Bar in 1890, and starting work on the South Wales Circuit. In 1906 he became Chairman of the Glamorgan Quarter Sessions before acting as recorder of the city from 1922 until 1930. He also entered politics after the Great War, serving as the Liberal MP for Banbury from 1918 until 1922, being appointed parliamentary secretary to the Minister of Transport in 1919 but only lasted two months in that post after falling out with the Minister. He later switched his political allegiance to the Conservatives and, after standing down from his legal and parliamentary duties, he spent plenty of time playing golf, both with his friends from the Club as well as at the newly-created Llantrisant and Pontyclun Club, of which he was the first President in 1927.

During 1943 Sir Robert Webber and his close circle of friends within the Club became involved in various fundraising activities for Eddie Price, the owner of a haulage contractors who had sustained serious injuries while working for the Ministry of Transport when a lathe fell off one of his lorries and pinned him to the ground. After spending ten days unconscious in Cardiff Royal Infirmary, Eddie came round and started to recover. Sir Robert and many of his pals spent time gathered around Eddie's bed and decided to help

> CARDIFF ARMS PARK.   Saturday, 13th January, 1945.
>
> "REMEMBER NOW the forgotten MEN IN BURMA"
>
> SEMI-INTERNATIONAL RUGBY MATCH ➡ GIVE ALL YOU CAN
>
> ## South Wales v. Sir Robert Webber's XV
>
> Kick-off 3.15 p.m.

their chum by providing him with a wireless radio so that he could keep up to date with events and listen to some of the wartime entertainment.

Eddie's radio became quite popular on the ward, so Sir Robert and his friends dipped into their pockets to provide other patients with similar equipment and headsets. Their generosity was warmly received so the group met up one day and gathered around Eddie's bed to discuss how they could expand their operations by raising funds to assist other injured servicemen. One of their first tasks was to give themselves a name; one of them said "look, there's nine of us around this bed, and it looks like the old so-and-so is going to get better so why not call ourselves Ten-of-us."

Within a few months, the group had started to plan the creation of a social club near Maindy Barracks where servicemen could relax, have a drink and forget the horrors of war. As part of the appeal, Sir Robert helped to arrange a rugby match on the Arms Park during January 1945 between a South Wales XV and a team bearing his name, and including many sportsmen from other parts of the UK who were stationed in the Cardiff area. The fundraising match raised both morale and some handsome funds, allowing the creation of the drop-in centre known as Cardiff House, as well as laying the foundations of Tenovus, which has subsequently become a leading cancer charity, supporting scientists prominent in many aspects of cancer research and care. Indeed, several members of the Club, including Professor John Lazarus and Fiona Peel – the Club's first lady member – have been hard-working members of the organisation's committee.

Returning to matters at the Clubhouse, by May 1943 Ivor Downing and the rest of the committee had become concerned about the state of the facilities, with the blackout curtains badly in need of repair and the furniture looking shabby. It was impossible to secure repairs or replacements, because of a shortage of supplies and specialist staff, so the officials created a reserve fund so that renewals could swiftly be sought once hostilities were over.

For several years most of the rooms in the Club had also been unheated, as coal had been in increasingly short supply. At first, log fires were lit for limited periods but, when firewood became as hard to find as the staff required to light and maintain the fires, the rooms were left unheated. A brief respite came during the autumn of 1943 when the Council's electricity department agreed to provide a heating circuit at a cost of £20 for the card room, but the books in the reading room had by now fallen victim to the cold damp conditions and were removed for pulping and recycling.

The following February, the Club was contacted about the removal by the military authorities of the metal railings around the front of the building. Like park gates and other railings across the country, they were requisitioned for ammunition and other purposes. Shortly afterwards, a short length of metal work in front of the card room was removed. Downing and the committee were concerned that more would be taken in 1944, but the authorities did not return.

Sadly, 1944 saw several Club members and their descendants lose their lives during the various operations associated with Operation Overlord on June 6th, as British and American troops invaded Northern France. From their newly-established bases, a decisive phase of manoeuvres began which, in addition to the retreat of German troops, hastened the end of the war. Maurice Turnbull was among those who was involved in the actions in Normandy but, tragically, he was to pay the ultimate price, as the all-round sportsman, now a Major in the Welsh Guards, lost his life in Montchamp in early August, 1944.

Maurice Turnbull, with his wife Elizabeth.

He had gone across with his battalion on D-Day +12, sailing from Newhaven for a night time crossing to Arromanches. His battalion advanced through the Normandy countryside, past bomb-damaged buildings and the bloated carcasses of farm animals, as well as human remains and the burnt-out military vehicles from the fierce battles that had taken place during the preceding days.

By the end of the month, the Welsh Guards had crossed the Orne River and across the Caen Plain where they became involved in Operation Goodwood, which has subsequently become regarded by military historians as one of the most complex and controversial of the Allied operations carried out in Normandy. Opinion is, in fact, still divided whether the aim was for a large British-led push on the eastern flank towards Caen, or if the British troops were being used as decoys to tie down the Panzer divisions in the east, thereby allowing the Americans to make a huge break out from Cherbourg in the west. Whichever point of view is taken, the effect of Operation Goodwood was that it forced the Germans to concentrate supplying weapons and fuel to their troops near Caen.

Maurice Turnbull (back right, standing) discusses matters in the Clubhouse with fellow members.

The assault on Caen began on July 16th and 17th, as two thousand British and American bombers dropped 8,000 tons of bombs in a 70 square mile area. During the manoeuvres, Maurice led No.2 Company in an attack on the village of Le Poirier where they met only light opposition, and took several prisoners, before successfully clearing all of the buildings. They remained there for ten days, by which time the assault on Caen was completed before moving west to clear any remaining pockets of German resistance and support the advancing American forces.

The area they moved into is known as the *Bocage*; a chessboard-like sequence of small fields, boxed in by hedged and sunken lanes, with a myriad of small villages, hamlets and woods. It was very dangerous territory in which to move, with plenty of hiding places from which the Nazis could mount a surprise attack. To make matters worse, the weather was foul, with prolonged downpours of rain turning the fields and their yellow-tinged soils into a muddy quagmire. These dreadful conditions, and the heavy German mortar fire and delays in the arrival of reinforcements, inflicted heavy casualties on the Welsh Guards as they attempted to move west.

By July 31st, the 1st Battalion had reached the hamlet of Pont d`Eloy, situated on a ridge some two miles away from the small town of Montchamp – an important communication centre and supply base for the retreating German troops. There were still delays with the arrival of American troops, so instructions were received for the Welsh Guards together with the 44th

The last known photograph of Major Maurice Turnbull (left).

Brigade of the 15th Scottish Division and a squadron of Coldstream Guards, to mount an assault on Montchamp to secure this strategically important centre.

The assault began in the early hours of August 4th, with Maurice leading No. 2 Company in a series of skirmishes which led to the Guards successfully capturing the town by the following day. Their brisk headway, however, had taken the Battalion commanders by surprise, and the tanks that were needed to permanently secure Montchamp still lay to the north of the town. Unbeknown to Maurice and his colleagues, there was also a strong pocket of German troops from the 9th S.S. Panzer Division to the south of Montchamp and on August 5th, they mounted a furious counter-attack in a bid to regain the town.

Maurice was supervising reconnaissance in the fields and orchards on the outskirts when the German counter-attack began with a column of tanks with foot soldiers either side and behind. Without any decent cover or supporting anti-tank weapons, Maurice told his men to hide alongside the hedge lining one of the orchards. As the tanks came closer, Maurice tried to send a message back to their headquarters, but their wireless communication had broken down, so Maurice told his number two to quickly run back into the town with news of the German attack.

Given the narrowness of the lane, Maurice knew that the advancing column of Panzers could be halted in their tracks if the lead vehicle was immobilised. He quickly assembled a group armed with guns and grenades and then, together with his men, crawled along behind the hedge in a bid to cripple the first tank. But, as they were alongside the Panzer and about to attack, machine-gun fire opened up and the tank's gun turret swung around, pushed through the hedge and opened fire. Maurice was right alongside and was hit in the back of his head, killing him instantly. Others in the Company were also either killed or wounded, before a hasty retreat began.

Maurice's body was later recovered and laid to rest in the British War Cemetery in Bayeux. News of his death reached Cardiff the next day. Grim notices were posted on the notice board both at the Cardiff and County Club and also in the pavilion at the Arms Park where Glamorgan, by strange coincidence, were playing a fundraising game in aid of the war effort. The crowd later stood in a minute's silence in Maurice's memory and, later that evening in the Club, glasses were raised in memory of a committee man who was destined for higher honours within the world of sport. Indeed, his high standing within the MCC had led him to be considered as the next secretary of the august organisation based at Lord's, but the tragic events in Normandy meant that this was not to be.

17

The Club's Garden and What to Do With It?

"It's a proper sort of place – an oasis where you can peacefully escape from the hurly-burly of life in Cardiff."

The news on May 8th, 1945 that the war in Europe was finally over was greeted with great delight. Across Cardiff people took part in spontaneous parties, and bunting and flags were placed across many buildings. The Cardiff and County Club was no exception as many members made their way into town to take part in great celebrations which went on, judging by surviving records, for 48 hours, after which Ivor Downing paid all the staff double wages for their efforts as well as granting them leave for the whole of the following weekend to fully recover.

Even though petrol and food rationing continued into the 1950s, life slowly started to get back to normal in the Club during the late summer of 1945. Downing and the committee set about the task of first sprucing up the furniture and décor, besides removing the blackout blinds and replacing the windows that had been damaged by wartime bombing. Once these housekeeping tasks had been achieved, they set about recruiting a new full-time Secretary at a salary of £300 per annum. Their choice was Morton Danby who took up his new position on February 1st, 1946. However, by early August 1948 he had resigned, with George Penn and his wife taking over the duties of Secretary and Manageress.

It was not long, though, before the committee were yet again dealing with the question of car parking, as the use of the garden became a major issue for discussion. With petrol having been strictly rationed there had been few grumbles about parking during the war, although in November 1944 the committee had noted wilful damage done to several cars which members had parked in the access road leading into the Arms Park.

During October 1946 the Club was contacted by the Secretary of the Cardiff Arms Park Company, who confirmed their intention to discontinue the

An aerial view of the Arms Park highlighting the size of the Club's garden.

gentlemen's agreement which Harry Cousins had brokered, giving members parking rights on the access road. The company was looking to let the roadway to car park contractors who would then develop a public, ticket-paying car park. A special committee meeting was swiftly convened, at which it was decided to convert the garden into a car park, with Sir Percy Thomas invited to advise on a suitable layout. But at a meeting on January 10th, 1947 the Vice-President, Sir Robert Webber, stated that many members were still opposed to doing away with the garden.

Shortly afterwards, Colonel Green, the Club's Honorary Solicitor, told the Club's officials that the Cardiff Arms Park Company was prepared to lease a portion of the road at a rent of £50 per year which formally gave the Club's members a right of way across the access road. The offer was swiftly rejected and the committee reaffirmed their intention to convert the garden, and instructed Sir Percy Thomas to prepare plans in advance of the forthcoming AGM in May. However, the anti-conversion lobby increased its support base and, at the Annual Meeting on May 10th, 1947, it was agreed that the garden should not be turned into a car park and that negotiations should be re-opened with the Cardiff Arms Park Company about the use of the access road and having a series of dedicated parking places.

Discussions duly began, but complaints still came in about vehicles blocking the access road, as well as passers-by urinating against the vehicles parked closest to the Club's kitchens. The Club paid £9 to the Council for the creation of a proper crossing for the access road, but problems still occurred with unauthorized vehicles blocking the entrance. The Cardiff Arms Park Company were so fed up about the situation that they contacted the Club to say that they would be looking to erect a series of gates across the entrance into Westgate Street. To some, this looked like the thin end of the wedge of barring Club members from using the access road and, even if the garden were converted at some time in the future, there would be a need for the Club to retain its right of way along the road and to the rear of the Clubhouse.

Sir Robert Webber, as President, led the anti-gates lobby and, after lengthy negotiations, the matter was finally resolved as the Club's Trustees and the Cardiff Arms Park Company signed an agreement dated February 17th, 1950 whereby the company erected a pair of gates across the Westgate Street entrance which were to be closed other than for use by authorized persons. Parking was also to be allowed on the side of the road next to the Club on weekdays, as well as on Saturdays when free passage was not required for matches at the Arms Park. Club members were also responsible for keeping the gates locked except between 12.30pm and 3pm.

Another issue which tasked the minds of the committee in the immediate post-war years was the question of holding dances, when ladies would be allowed into the Club and invited to partake in the socialising. Balls had been held for many years in the Town Hall, and subsequently the City Hall, whilst the Club garden had hosted a number of parties at which wives and daughters and other female friends had joined members to relax and unwind. Allowing women inside the premises was a very different matter and, in the eyes of some, a daring proposition so in November 1947 a referendum was held. The result of the ballot, 70 votes in favour and 30 against, saw ladies being admitted to the Club over the Christmas period for a dance evening. Despite losing the vote, the dissenters appear to have been of sufficient influence that no dance was arranged in December 1947.

However, on April 21st, 1950 a dance was held, with Geoffrey Howard's six-piece orchestra being booked for the sum of £80 16s to play for five hours during a party which ran from 9pm until 2am.

The Club's garden is flooded by storms in September 1961.

The Suggestions Book

Over the years, this document has prompted some interesting comments, such as the one below: submitted at a time of austerity during the Second World War.

Date.	Suggestion.
1942 June 25.	That members shall be permitted to introduce a Guest – who is not a relation or an in-law – or a Distinguished Person, at any time of the day (with the exception of the mid-day meal on week-days) and such Guests will only be allowed to share the member's ration of Alcoholic drinks.

[Signatures of members]

175 tickets were sold at 35 shillings each, with a head waiter and chef from the Queens Hotel also being employed to supply the buffet which was available in the dining room. The dancing itself took place in the billiard room upstairs and the evening was regarded as such a success that it was repeated the following year on April 13th, 1951.

In 1953 a cocktail party was also held in the Club to celebrate the Queen's Coronation and for the first time the female guests were able to use the upstairs rooms. Several though were appalled by the state of décor. 'Those curtains are in ribbons' and 'Why is this tatty place so special?' were just two of the comments directed at the Club's staff, who soon afterwards took steps to rectify matters by redecorating these areas.

These, more contemporary, events reflected the progressive mood within Cardiff and much of south Wales in the years immediately after the Second World War. The city had already benefitted from the gift in 1947 by the Marquess of Bute of all of his land, including the Castle, with Bute Park, Sophia Gardens, Pontcanna Fields and other areas of open space, which were to be preserved as recreational land. There was further good news on December 21st, 1955 when the Lord Mayor received formal confirmation from the Home Secretary that by royal decree of Her Majesty the Queen, Cardiff had attained the privileged position as capital of Wales.

The news was formally conveyed the following day at a special ceremony with Cennydd Traherne, a member of the Club since 1933 and the Lord Lieutenant of Glamorgan since 1952, making the announcement to the delight of the aldermen and other councillors who gathered in front of the City Hall. Born in 1910 at Coedarhydyglyn, the elegant Regency mansion near St. Nicholas, Traherne had grown up on the family's spacious estate with its magnificent views over the Ely Valley, before attending Wellington College and reading law at Brasenose College, Oxford. He had also taken part in Operation Overlord, landing on the Normandy beachheads on D-Day +1, before being mentioned in despatches when serving with the 102 Provost Company of the Royal Military Police.

He had switched a year before from the 81st Field Regiment of the Royal Artillery but, after a few training ground mishaps, it was clear to him that his future did not lie as a gunner. It proved a wise move as he became Deputy Assistant Provost Marshall of the Second British Army in 1945, and served with the Royal Military Police until 1949. During the war,

Sir Cennydd Traherne.

his devoted wife Rowena had also been a delivery pilot for the RAF, flying Spitfires and Hurricanes from various factories to the military airfields.

Before the war, Cennydd Traherne had practised as a barrister in London and, after leaving the Royal Military Police, worked in south Wales, serving as deputy chairman of the Glamorgan Quarter Sessions from 1949 until 1952, besides looking after the affairs of the family's estate. He also keenly promoted the city of Cardiff as various officials considered, from 1951, the merits of attracting major developments to Wales and the city's claim to be declared the nation's capital.

There had been good news the year before as the organising committee of the British Empire and Commonwealth Games announced that Cardiff would host the 1958 event. A delegation from Cardiff duly travelled to Vancouver in British Columbia for the 1954 Games when the formal announcement was made. Soon after their return, city officials began planning for the sporting event, including the construction of a new swimming pool, fittingly called the Empire Pool, in Wood Street. The Pool, built at a cost of £700,000,

The Clubhouse, as seen in the early 1960s.

had seating for 2,400 spectators and there was a full house to watch the swimming and diving events held during the 1958 Games.

The Pool was opened on April 18th, 1958 by the Lord Mayor, the former *Western Mail* sports journalist and Club member, John Hinds Morgan, who had also been the last person associated with the Club to see and speak to Maurice Turnbull, bumping into his former Club acquaintance in Normandy a few days before his death. Morgan later wrote about this remarkable meeting: 'I was in a Jeep moving up nearer to the fighting, but in the congestion of that Normandy lane, we had to pull in to the side to allow the Guards to pass through. Familiar Welsh voices brought a reminder of home and then I spotted Maurice – the same upright, alert figure, but garbed in battle-dress instead of the white flannels of the cricket field. I promptly jumped off the Jeep and so far as military discipline would allow walked by his side. 'Fancy meeting you here,' he said. We hardly knew what to talk about and I was fearful of breaking Army etiquette. There were just a few personal exchanges, and then he was gone.'

There was plenty of involvement of Club personnel in the Games, as Maynard Jenour, a long-standing member and a President in later years, was deputy organising chairman for the Games. Cennydd Traherne, in his capacity as Lord Lieutenant of Glamorgan, was also host to the Queen during her visit and he was in the royal party when, during the Cardiff Games, she formally announced that Prince Charles was to become the Prince of Wales.

The Games themselves were staged between July 18th and 26th with the Arms Park hosting the athletics as well as the opening ceremony. The cycling was held at Maindy Stadium, with the 120-mile road race taking place around a seven-and-a-half mile circuit near the River Ogmore. The boxing and wrestling took place in the newly created Pavilion in Sophia Gardens, using a converted RAF hangar, and Cae'r Castell School in Llanrumney was used for the fencing. The Cardiff, Mackintosh and Penylan clubs all hosted the lawn bowls competition, with the weightlifting being staged at the Barry Memorial Hall and the rowing at Llyn Padarn in Snowdonia in north Wales.

The Arms Park, though, was the epicentre of the Games with the Club and its ever-growing list of members being ideally situated to take a grandstand seat. Indeed, for many months before the event, the committee painstakingly discussed the arrangements for the Club's members to watch and enjoy the spectacle of the opening ceremony, although it was only at a late stage, and just a few weeks before the games began, that the athletics were switched from Maindy Stadium after concerns had been expressed about the capacity and facilities at the track to the north of the city centre.

It was, therefore, fortuitous that the Club's administrators had agreed to the creation of an area of seating, stretching for 70 feet along the roof of the Clubhouse. Under the guidance of architect Sir Percy Thomas, a two-tier bank with 86 seats was safely installed. Elsewhere in and around the Club, a commissionaire was engaged for the duration of the Games, and was on duty from 12.30pm until 7.30pm each day to meet and greet the many members who were likely to be in town with a number of guests. Many dignitaries dined as guests, including officers from HM Yacht Britannia and HMS Orwell which accompanied the royal vessel. In August 1953, His Royal Highness The Duke of Edinburgh had also consented to become the Patron of the Club, although there are no records of him dining at the Club during the Games.

A set cold lunch was also served each day, with a tea in the dining room from 4pm until 5.30pm, but each were solely for the men folk, with ladies being able to have tea in the withdrawing room on the ground floor next to the entrance into the garden. The Club's car park was closed throughout

The 1958 Empire Games attracted huge crowds to Cardiff city centre.

The Club's Garden and What to Do With It?

with the much-talked about access road being used by the marathon runners as they made their way out of the Arms Park and into Westgate Street before their race around the streets of the city and its surrounding areas. Fortunately, there were no vehicles blocking their way as they ran out from the stadium!

Competitors in the marathon at the 1958 Empire Games leave the Arms Park and turn into Westgate Street.

18

Sir Tasker and 'The Skipper'

"A refuge, full of tradition and respectability."

The recruitment drive during the 1950s saw a number of new members joining the Cardiff and County Club, with a host of prominent figures in Welsh life being admitted. Among those admitted in 1958 was Tasker Watkins, a leading figure in the legal world who had won the Victoria Cross after a series of heroic acts in Northern France during the weeks after the D-Day landings.

Born in Nelson and educated at Pontypridd Boys' Grammar School, Tasker initially taught in London before joining the Army in 1939 and serving in the ranks before being commissioned into the Welch Regiment on May 17th, 1941 as a second lieutenant. Like Maurice Turnbull, he was also involved in operations in Northern France, clearing away German troops following the Normandy landings.

In particular, on August 16th, 1944, Watkins won the VC after his battalion had been ordered to attack objectives near the railway at Bafour, as part of a move to trap the Fifth and Seventh German Armies in the vicinity of the nearby town, known as the 'Falaise pocket'. To achieve their objective, Watkins and his company had to cross a series of cornfields containing a number of booby traps, and all while under heavy machine-gun fire from German positions elsewhere in the fields. The mounting number of casualties slowed the advance of the troops and Lieutenant Watkins found himself the only officer remaining. He put himself at the head of the surviving soldiers and, despite the short-range fire from the Germans, he and his men charged two enemy posts in turn.

On reaching his objective, Watkins found an anti-tank gun manned by a German soldier. At that vital moment Watkins' Sten gun jammed, so he threw it into the German's face and shot him with his pistol. A group of 50 or so Germans then led a counter-attack but, despite being down to just 30 men, Watkins led another bayonet charge which repulsed the enemy.

Having lost their radio communications with the rest of the Allied forces, Watkins then led his company back to the rest of the battalion by moving around the flanks of the enemy positions in the cornfield. Once again, he was challenged by an enemy post at close range and, after ordering his men to scatter, Watkins charged the post with a Bren gun and silenced it before leading the remnants of his company safely back to battalion headquarters.

Watkins' citation recorded that 'his superb gallantry and total disregard for his own safety during an extremely difficult period were responsible for saving the lives of his men and had a decisive influence on the course of the battle', which resulted in the capture of 50,000 German prisoners. He was promoted from lieutenant to major in the field, and subsequently was decorated with the Victoria Cross by King George VI at Buckingham Palace on March 8th, 1945, and so became the first Welshman serving in the Army to be awarded this honour.

The bravery of Sir Tasker

Reflecting some 50 years after his heroic deeds, Sir Tasker Watkins said: "You must believe me when I say it was just another day in the life of a soldier….I did what needed doing to help colleagues and friends, just as others looked out for me during the fighting that summer…. I didn't wake up the next day a better or braver person, just different. I'd seen more killing and death in 24 hours – indeed been part of that terrible process – than is right for anybody. From that point onwards I have tried to take a more caring view of my fellow human beings, and that, of course, always includes your opponent, whether it be in war, sport or just life generally."

De-mobbed in 1946, he subsequently read for the Bar, and was called by Middle Temple in 1948 before starting to practise in common law on the Wales and Chester Circuit. He was deputy counsel to the Attorney-General, Sir Elwyn Jones, at the Aberfan Disaster Inquiry in 1966, besides prosecuting in a number of high-profile cases involving Welsh extremists, including the Free Wales Army trial in 1969, which followed the discovery of a plot to attack Caernarfon Castle and assassinate Prince Charles.

Knighted in 1971, he subsequently headed a working party set up by Lord Chief Justice Lane which proposed changes in Crown Court procedure designed to speed up and cut the costs of criminal trials. His appointment in 1982 as the first Senior Presiding Judge for England and Wales was designed to relieve Lane of some of his heavy administrative burden. In 1988 Watkins was appointed Deputy Chief Justice of England and in 1991 sat alongside Lord Lane in the historic appeal case that established that husbands living with their wives can be convicted of raping them.

A straightforward and formidable man who held strong views on many matters, Sir Tasker had a heart of gold and was great company in the Club, especially on Saturday mornings during the winter months when he would regularly call in, and have three halves before heading west and watching Glamorgan Wanderers play at the Memorial Ground in Ely. In 1993 he retired from the bench, aged 75, and was also voted in as the new President of the Welsh Rugby Union, a position he held until 2004. He was among

Wilf Wooller (third right) is introduced to the Prince of Wales at Twickenham before the start of the England v Wales rugby international in February 1933.

the guests at a number of special functions in the Club, including being guest of honour at a special celebration in 1971 held to commemorate his knighthood.

1957 saw Wilf Wooller being admitted to the Club, almost a decade after he had led Glamorgan to their first County Championship title, and more than two decades since his finest moments on the rugby field for Wales. Indeed, rugby was his first love and, when at Rydal School preparing for his entrance exams to Cambridge, he formed a fine partnership in the centre for Sale RFC with established Welsh international Claude Davey which swiftly saw Wooller elevated to the Welsh side.

In February 1933 he was alongside Maurice Turnbull in the Welsh three-quarter line when the men in red won for the first-ever time at Twickenham. While reading Anthropology at Cambridge, he won further Welsh caps, besides rugby Blues between1933 and 1935, plus cricket Blues in 1935 and 1936. He also played a key hand in Wales' historic 13-12 victory over New Zealand at the Arms Park. It was a game in which the home side had been reduced to 14 men because of injury as those were the days before replacements were permitted. Wooller's shrewd kicking led to two tries which sealed an historic victory and yet another night of merriment and joy in the Club.

After coming down from Cambridge, he accepted a post in the coal trade in Cardiff and played rugby for the city club, besides playing cricket for St. Fagans. The latter led to both a whirlwind romance and marriage to the daughter of the Earl of Plymouth, on whose estate the St. Fagans side play, plus selection for Glamorgan for whom he appeared while on leave from his duties at Cardiff Docks. Among the highlights of his games before the war for Glamorgan was his match-winning century in the victory over the 1939 West Indians at the Arms Park. Wilf also guested

Wilf Wooller (left) at a function with Jack Manchester, the former All Black, and Viv Jenkins.

for Cardiff City FC in some of their exhibition matches, and played squash alongside Maurice Turnbull and some of the city's other talented sportsmen at the city's newly-established squash club in Ryder Street in Riverside.

When war was declared, Wilf and many of his friends from Cardiff's sporting fraternity, enlisted with the 77th Heavy Anti-Aircraft Regiment, with Wilf undergoing training as a gunner. He was subsequently posted to the Far East and, in Java during 1942, was captured by the Japanese. His war was over and for the next three years as a prisoner of war he was incarcerated in Singapore's notorious Changi prison. Back home, he was listed as missing and, for a year at least, it was presumed that he was missing in action. For a while, Wilf worked on the infamous 'Death Railway' in Burma, a hardship to which many others succumbed. The experience certainly left a deep mark as, years later, he flatly refused to use a Japanese-made pocket calculator or other pieces of equipment made in the Far East, and he often woke up in the middle of the night, screaming and uttering oaths after having nightmares about these terrible years.

After returning to south Wales as little more than skin and bones, he was unable to play rugby again but, through a long-standing friendship with Johnnie Clay, he took an increasing interest in Glamorgan CCC. Instead of being a glorified hobby, cricket became his main interest and in 1947 he took over as the Glamorgan captain, before the following summer leading them to the County Championship title for the first time in their history. During the summer of 1948, as well as many subsequent years, Wilf took great delight in being invited by Johnnie, together with others, to dine and drink after play in the Club, and like Maurice Turnbull before them, engage in lively chat with the visiting amateurs.

If Wilf's finest hour on the cricket field in 1948 came at Bournemouth with the defeat of Hampshire to clinch the Championship title, the greatest personal moment for the man known to his team as 'The Skipper' also came that year with his marriage to Enid James. After the deep sadness at the dissolving of his first marriage to Gillian Windsor-Clive, Wilf found a true and loyal soul mate in Enid, the pair having five children, with Jon following in his father's footsteps by becoming active and popular member of the Club.

He continued to play for Glamorgan, besides acting as their Secretary until 1960. There was a brief return for a solitary match in 1962, but Wilf then switched his attention to the secretary's duties, besides being a journalist with *The Sunday Telegraph* as well as with BBC Wales, and also working as an insurance broker. As a broadcaster and sports writer, Wooller was on several occasions quite outspoken. He had always been a strong-minded figure on the field, freely giving advice to opposing batsmen while fielding at short-leg and snarling at them. Occasionally his written words gave offence, and the mother of a Welsh rugby international sent him a pair of spectacles through the post after Wooller had criticised her son for being caught out of position. The glasses though were returned with a covering note which said that as his wife was an optician, he had been assured that he could see perfectly well!

During the 1960s and into the 1970s, Wilf's uncompromising support for keeping sporting links with South Africa when *apartheid* was at its most intense sat uneasily with many other people who believed that sanctions should mean no sporting, cultural or trading links whatsoever with South Africa. He had several heated public debates with Peter Hain, who had led opposition to the involvement of South African rugby teams in Britain. It seemed that nothing excited Wilf more than debating controversial views which he duly defended as ardently as he had played cricket for Glamorgan or rugby for Wales.

Wilf could therefore be a lively and stimulating guest in the Club, and a host of members enjoyed his company. One of these was his old friend Arthur Rees, who had become a member of the Club in 1970. Born in the village of Llangadog in 1912, Arthur was raised at first as a Welsh speaker, and did not learn English until the age of seven. Educated at Llandovery College, he was also up at Cambridge during the 1930s, having gained a place at St. Catharine's College. He won 13 caps as a back-row forward in the Welsh rugby side between 1934 and 1938 as well as a pair of rugby Blues, and Arthur was alongside Wilf and fellow Light Blue, Cliff Jones, in the Welsh side which defeated the All Blacks at the Arms Park in 1935 – a match which elevated them from callow youths into national heroes.

After graduating, Arthur joined the Metropolitan Police and then the Royal Air Force after the outbreak of the Second World War, where he rose to the rank of Squadron Leader before ending as acting Wing Commander. After the war, he returned to the Metropolitan Police and subsequently became Chief Constable of Denbighshire in 1957, followed in 1964 by becoming the Chief Constable of Staffordshire. He also served on a number of sporting bodies, including the Midlands Sports Advisory Council as well as the British Karate Board, besides being a director of Stoke City AFC. In May 1977 he was granted the Freedom of the City of London.

He subsequently moved back to south Wales and there were many occasions in the Club when Wilf and Arthur would recall their days as undergraduates and the fun they had shared while at Cambridge. By a strange quirk of fate, the pair shared the same birthday and as students became known as the 'Twins'. Their success on the rugby field also meant that they were treated like gods by the other young Welsh scholars who were at Cambridge, with the pair often being mobbed by fellow students as they made their way around the university.

On occasions, they also attracted the wrong sort of attention, most notably on one afternoon in a highly respectable restaurant in Cambridge. Wilf and Arthur had gone into the Dorothy Café for tea, crumpets and a quiet chat, and were minding their own business when in walked the tall and bulky figure of the university's heavyweight boxing champion. He was clearly out to impress some of his female companions so, after recognising the pair of rugby players, he walked over and purposely bumped into their table, spilling the teapot and sending the crumpets and other cakes flying on to the floor. The boxer, who clearly had more brawn than brain, burst out laughing, but Wilf got up, calmly looked him in the eye and, with a short arm jab to his chin, sent him to the floor where he lay dazed for several minutes. In later years, the boxer became the bodyguard to the leader of the British Union of Fascists, Sir Oswald Mosley.

Fortunately, nobody bumped into Wilf's or Arthur's table when they were dining or chatting with friends in the Cardiff and County Club, and the pair were great company and raconteurs, each with a fund of humorous tales about the carefree days of amateur sport during the 1930s.

Wilf Wooller practising his rugby skills in a Cardiff park during the 1930s.

19

The Beeb

"A place for the opinion-formers of local life."

February 13th, 1923 had seen the first BBC broadcast from a studio in Cardiff as, at 5pm precisely, a radio programme was transmitted with Mostyn Thomas singing *Dafydd y Garreg Wen*. In July 1937, Sir John Reith, the BBC's director general, visited Cardiff for the inauguration ceremony of what was called 'The Welsh Region' with studios in Swansea, Bangor and other places in Wales being added to the range of locations from where current affairs, news and sport were broadcast. A number of rooms across Cardiff doubled up as studios, with premises as well in Park Place and Newport Road, but these were all consolidated in 1966 by the opening of Broadcasting House in Llandaff. This followed the launch in February 1964 of BBC Cymru Wales, and new transmitters across the country, as well as on the city's edge at Wenvoe, all improved the national coverage.

The 1980s saw S4C established as a Welsh fourth channel, and in 2005 *Doctor Who* was re-launched by BBC Cymru Wales, having not been seen on British TV screens since 1989. Under the guiding hand of executive producer Russell T. Davies and BBC Wales' head of drama Julie Gardner, the series proved to be an instant hit and has subsequently been a flagship production for BBC Wales with a host of locations around Cardiff and its suburbs being used as the setting for scenes from the sci-fi series, as well as its spin-off programme *Torchwood*.

Recent investments by the corporation, such as the Roath Lock studios and the creation of the Doctor Who Experience in Cardiff Bay, as well as the building of a new Welsh Broadcasting House in the Central Square development in the city centre, all illustrate the importance of the media industry in the modern post-industrial economy. But even in the age before digital TV and broadcasts in colour, BBC Wales played a key role in the local economy, with a string of famous broadcasters and producers all being prominent members of the Cardiff and County Club. Pre-eminent among these was G.V. Wynne-Jones, the popular sports commentator, known to one and all as Geevers.

Born Griffith Vernon Wynne Jones (without the hyphen) in Corwen, Meirionethshire, in September 1910, Geevers was educated at Cardiff High School, and subsequently Christ College, Brecon. A member of the Club since July 1950, he became a highly successful journalist after the Second World War, and familiar voice on BBC Radio. He was good friends with Freddie Mathias, with the two men sharing a passion for aeronautics and, when Wilf Wooller also joined them at the dining table in the Club, the conversation and laughter would soon flow.

Geevers had played as three-quarter for Bristol, before moving across the Severn and playing for both Cardiff and Newport. His move to south Wales also saw him get married in October 1935 at Llandaff Cathedral to Rhona, the daughter of Fred Pipe, a fine batsman with Cardiff CC and a man who, through his regular and well-paid work at the Bute Docks, had to turn down opportunities to play for Glamorgan as a professional cricketer.

After the Second World War, Geevers became the voice of Welsh rugby, as he was part of a group of pioneering sports broadcasters on BBC Radio, as well as later the BBC Welsh Home Service, and BBC Radio 4 Wales. His commentary position, though, at the Arms Park was not as luxurious as those in the modern stadium, with Geevers and his intrepid colleagues having to use the greyhound racing judge's box, situated on top of the South Stand, and accessed by a trap-door in the roof after climbing up a ladder.

He was very much a champion of Corinthian values and all that is best in amateur sport and, during 1951, he was involved in a spat with the Welsh Rugby Union after he had threatened to name the clubs who he alleged had been paying players and who were paying excessive personal expenses. Things reached a head during March when the Union refused his admission into the Arms Park to broadcast the Wales-Ireland international, with another commentator having to take his place. The ban remained in place for a year, but Geevers stuck to his guns and did not back down.

Geevers had several business interests, working initially as sales director for GKN. At times when steel was in high demand, his job was quite straightforward but, as demand dropped, he found it harder and harder, and he duly left the company in 1959. He invested his severance money in running a holiday business in Pembrokeshire and having a brief flirtation with the world of music. In the case of the latter, he helped to set up a record business called Qualiton and, during the late 1960s, employed a young girl called Mary Hopkin on one of the disc-cutting machines. As he later said, in his charming and self-effacing way, "Mary didn't come into work one day and, after making enquiries, I found that she had gone off to be a singer. Nobody had told me she could sing!" Indeed, the next time Geevers saw Mary was on television after she had achieved fame with *Those were the Days*, which topped the UK singles charts in 1968.

The story of how close he had been to signing a pop star was just one of the many stories Geevers loved to relate to his friends in the Club and, with his vast experience of the worlds of industry and sport, an evening in his company was great fun. Indeed, as one of his close friends once related, "to have supper with Geevers was to sit next to a vast tub of knowledge. You dipped your question in, and out came the answer, whether it was on sport, politics, business or merely the weather. And it was never simply information – it was always with a story."

Geevers also took great delight in speaking to any of the barristers staying in the rooms on the upper floors during the Assizes week. Rather than having a quiet evening relaxing and dining in the Club, they would face an evening of light-hearted cross-examination by Geevers, as he quizzed the visitors about the events of the day and the alleged crimes of the miscreants, before regaling his audience of new acquaintances with a series of whimsical tales.

He was also famous – some might say infamous – for a witty retort. One of his most famous asides came at a function in the Club when he and his drinking pals had the pleasure of the company of a new member, who was

Geevers (right), with Freddie Mathias (left) and the Lord Mayor of Cardiff (centre).

trying far too hard to impress and exaggerating his recreational exploits. When asked about his fishing and whether some of his new acquaintances could join him, the young chap puffed out his chest and replied "Yes, I'm sure I could fix you all up with some salmon fishing as I have access to one of the finest stretches on the River Wye, and yes, I could also help out with some deep sea-fishing as well as I have one of the latest power boats with five large berths." Geevers could not bear to hear these boastful remarks any longer, and interjected: "For God's sake, don't ask him about flying – he's got Concorde in his greenhouse!"

Over the years, Geevers also built up a handsome collection of souvenirs from many aspects of Welsh sport and these subsequently formed the basis of

Geevers and Lord Ted

During October 1964, G.V. Wynne-Jones enjoyed what can only be best described as some quite interesting weeks escorting Ted Dexter, who was standing as the Tory candidate against Jim Callaghan in the Cardiff South-East constituency in the General Election of October 1964. Apart from politics, it could barely be described as a meeting of kindred spirits, as the 29-year-old England cricket captain, who had opted out of the winter tour to South Africa, had an aristocratic air which had earned him the moniker 'Lord Ted', besides having few links, if any, with the constituency he was hoping to represent, especially in some of the working-class neighbourhoods of Cardiff.

In the minds of many, Dexter was a celebrity candidate, yet Geever's, as the candidate's ever-loyal minder, spent the weeks during the lead-up to the polls telling the good folk of Roath, Splott, Grangetown and the docks, as well as any doubters, that Ted was here on merit and not, perish the thought, because of his sporting prowess. He entertained Dexter several times at the Club, before Lord Ted duly opened his political innings in the Roath Conservative Club. He began by saying: "I was, until this evening, on a sort of trial. But Cardiff has given me my colours," before using a plethora of sporting metaphors and briefly reminding his would-be voters of his cricketing pedigree.

Later, he told an audience of dockers and steelworkers that he was one of them, but Jim Callaghan, defending a majority of only 868, ended up securing a majority of 8,000 as, to use Lord Ted's own analogies, the man batting for the Tories rather trod on his own stumps during the weeks leading up to the general election, much to Geever's private embarrassment.

the Welsh Sports Hall of Fame, something about which he had campaigned for many years and had discussed with his friends at the Club. Awarded the OBE in the Queen's Birthday Honours List in 1957, Geevers was for many years a pillar of Cardiff Athletic Club, as well as being a lynch-pin of Cardiff South-East Conservative Club, serving as chairman for 15 years and president for an additional six.

Geevers lived with his wife in the flat on the top floor of the Clubhouse and for several years combined his duties as the Club's secretary with those of kitchen manager. He was excellent at all of the administrative duties and there was probably no finer front of house or *bon viveur* in Cardiff. However, his work overseeing the hygienic operation of the kitchen was less effective and resulted in Geevers' temporary removal from those duties. His replacement, however, only lasted for a short period, with Geevers happy to return and willingly helped out the Club at a potentially difficult time.

Wynford Vaughan-Thomas was another well-known broadcaster who was a frequent visitor to the Clubhouse in Westgate Street. Born in Swansea in 1908, he enjoyed an illustrious career with the BBC and was most famous for broadcasting live from a Lancaster bomber during an RAF mission over Berlin during 1943.

He was the second son of Dr. David Vaughan-Thomas, a music teacher, and attended Swansea Grammar School at the same time as Dylan Thomas, and was taught English by Dylan Thomas' father, before reading history at Exeter College, Oxford. He then had a brief period in teaching, before securing a temporary post in 1933 at the National Library of Wales, as curator of manuscripts and records. After a stint as south Wales regional officer for the Social Services Council, in 1937 he joined the Outside Broadcast Department in Cardiff, largely so that he could be near his fianceé, Charlotte Rowlands, whom he married in 1946.

One of Wynford's first commentaries was during the coronation of King George VI in May 1937, when he delivered the broadcast entirely in Welsh, and conveyed what was happening around him in his typically vibrant way. Shortly after the outbreak of war, he moved to London and, during his time as a home front correspondent he reported on the Blitz. In 1943 Wynford accepted a post as a war correspondent and, during September that year, he and an engineer joined the crew of a Lancaster bomber during a raid on Berlin. He later admitted that the eight-hour flight was completely terrifying, and "beautifully horrible", but few listeners detected any fear in his voice as he described the bombing "as like watching somebody throwing jewellery on to black velvet, winching rubies, sparkling diamonds all coming up at you."

His further description of the Lancaster flying through the searchlights, and the terrifying moment when it was caught in one of the beams, encapsulated the dangers faced daily by the RAF crews, besides winning Wynford many admirers within the audience and the BBC's executives at Broadcasting House.

The following January he landed on the beaches of Anzio in Italy with British and American soldiers before reporting on the assault on Rome. During June and July he reported from Normandy after the D-Day landings, besides joining members of the French Army and was subsequently awarded the *Croix de Guerre* in 1945 for being with the soldiers who liberated the Burgundy vineyards. When asked later about these exploits, he replied in typically jocular fashion "We had three marvellous days in a cellar and I emerged with the *Croix de Guerre*!"

There was nothing to laugh about, however, later in May and June 1945 when he worked from the studio in Hamburg where the traitor, William Joyce, known as Lord Haw-Haw, had broadcast his Nazi propaganda to Britain. Wynford was also the first to broadcast from the Belsen concentration camp, graphically describing the horrors that surrounded him. At the end of the war he returned to London and became the Royal Family correspondent – a post he held for more than 20 years – and besides reporting on a number of state occasions, including the Royal Wedding in 1947 and the Coronation in 1953, Wynford accompanied the Royal Family on several overseas tours.

In 1967 Wynford became one of the founders of the commercial television company serving Wales, Harlech Television or HTV (which had, that year, won the Wales franchise from TWW), and three years later took up a post as the company's acting director. He moved back to Wales and became a more regular face in the Club, besides becoming director of the Welsh National Opera Company, and the president of the Council for the Protection of Rural Wales. He also wrote a number of books, some of which described his wartime experiences, as well as a number of volumes on the history of Wales and the Welsh landscape. Wynford was awarded the OBE in 1974, followed by the CBE in 1986. The following February, he died at his home in Fishguard.

For many years he had been a member of what was jokingly referred to as the Club's unofficial wine committee. Along with Chris Cory – one of the Chairmen during some of the difficult years in the 1970s – William Crawshay and Dennis Martyn, they regularly wined and dined in the Club, passing comment on the merits, or otherwise, of the various offerings. Chris was the brother of John Cory of St. Bride's and was very much a *bon vivant*, sharing with Wynford and his pals a love of good wine and in particular burgundy. Wynford spent many a long lunch, and evening, in the company of his good friends in the Club and, after being spotted by visitors, he would willingly recount some of his wartime stories or tales about life on a foreign tour with Her Majesty the Queen.

Wynford also had a puckish sense of humour and, together with his circle of friends in the Club, he would enjoy the after-dinner challenge of devising saucy limericks. He would readily accept a host of requests from fellows who thought that it would be impossible to find something that rhymed with a particularly unusual Welsh place-name. To the delight of his pals, Wynford – after a few minutes of thought and perhaps some inspiration from the glass in front of him – would always come up trumps, even when faced with the potentially tricky task of composing a limerick involving Blaenau Ffestiniog, as deftly achieved below:

> 'In the groves around Blaenau Ffestiniog
> The girls 'kiss' for a penny or ceiniog
> I know for a fact
> Because I was caught in the act
> By Lord Hailsham, known then as Quintin Hogg!'

Wynford Vaughan-Thomas, in his early days as a broadcaster.

20

Celebrating the Club's Centenary

"It's just the right size and has an informal atmosphere which other clubs just simply have not got."

The 1950s had seen further changes in Club personnel following the retirement of George Penn and his wife as Secretary and Manageress in 1950. Their immediate replacement was far less effective with record-keeping and, after the discovery of pilferage in the wine cellar and provision store, plus other irregularities with telephone charges and postage, there was a parting of the ways. Commander Edward Tipple took over as Secretary and Steward during the mid-1950s and, given his distinguished naval career, it would be fair to say that he put the ship back on even keel.

Like previous administrators, it wasn't long before the Commander was embroiled in further discussions over the future of the garden. There had been concern in various quarters about the number of parking spaces available and whether during busy periods such as international weekends and the forthcoming Empire Games, the existing provision was sufficient. Once again, the question of finding space in the garden was raised, and Sir Percy Thomas submitted a plan in October 1957 for a driveway around the centre lawn where a single line of 14 cars could be parked. The plan was rejected as the pro-garden lobby won the day.

Problems continued over the following months, especially with people passing themselves off as Club members and using the parking spaces. A joint meeting therefore took place with Cardiff Athletic Club and the Greyhound Racing Company, at which it was decided to issue identification badges for vehicles, while the Club would keep a register of all members' cars and their registration numbers. With the rise in car use, the two dozen parking bays on the one side of the access road was felt to be insufficient and, on January 9th, 1961 the Club secured the use, for an annual fee of £200, of parking spaces on the other side of the access road. It was also made clear to Club members that wives, friends and relatives could not use these spaces and that chauffeur-driven vehicles would only be permitted into the space if they had the member's car pass card and front door key. All of these arrangements would be overseen by a car park attendant who would be on site from noon until 3pm.

In August 1963 Commander Tipple handed over the Secretary's duties to Major Frederick Deamer, a retired Army officer who had entered the licensed trade and had run, with great effect, The Five Horse Shoes in Stamford, Lincolnshire. However, the major and his wife were in place for only a couple of years before they also parted company with the Club, leaving G.V. Wynne-Jones to fill the position of Secretary and to oversee the arrangements for the Centenary Dinner on February 1st, 1966. 'Tip' Williams had also succeeded Sir Robert Webber as President in 1962, and, together with the committee, they initially asked Wynne-Jones to make arrangements with The Angel Hotel.

CENTENARY DINNER

THE QUEEN
The President

THE CLUB
Sir Cennydd Traherne
Her Majesty's Lieutenant for the County of Glamorgan

THE PRESIDENT, Mr. L. E. W. Williams
will respond
supported by
THE CHAIRMAN OF THE COMMITTEE
Mr. E. R. K. Glover

Menu

Smoked Salmon
and
Potted Shrimps

Turtle Soup *Avery's Silvers*

Roast Aylesbury Duckling
Petits Pois *Chambolle Musigny Les Amoureuses 1959 (Avery)*
Duchesse Potatoes

Roes on Toast

Coffee *Sandeman 1945*

The menu card for the Centenary Dinner.

Toasts were made by Sir Cennydd Traherne, 'Tip' Williams and Ted Glover, the Chairman of the committee. The latter pair of gentlemen each had a fund of stories about life in the Club, as well as on the sports field, with each having played county cricket for Glamorgan, besides enjoying jolly times with the South Wales Hunts CC, and – admittedly with less success – on the golf course with the Club's Golf Society.

Several longstanding members questioned the use of The Angel Hotel when perfectly adequate dining facilities existed within the Club. Consequently, an emergency committee meeting was held on February 6th, 1965 at which it was agreed that the proper place to hold the Centenary Dinner was at the Club and a circular was sent to all members. After receiving replies which indicated around 165 might attend, plans were set in motion to accommodate them within the Club's premises. Hancock's were asked to assist in providing sufficient equipment and seating accommodation in the dining room, reading room and the television room, which had been created in one of the upstairs rooms.

The ticket price of three and half guineas, including dinner, wine and liqueurs, was agreed together with a seating plan so that members were seated in order of seniority in each of the three rooms. Loudspeakers were installed in each room so that the speeches could be conveyed to all rooms, while surplus furniture was taken away for 48 hours by a local Army unit. Additional cloakroom space was provided on the first floor, and an agreement was reached with the Greyhound Racing Company for extra car parking behind the eastern stand at the Arms Park.

A garden party was also planned for June 10th, 1966, with initial thoughts of having some 200 attendees including guests and ladies. Plans were set in motion to hire a three-piece band, and a piano, and thought was also given to a wet weather contingency plan. However, the event never took place and was replaced instead by a cocktail party from 6.30pm until 8.30pm.

By the time the members of the Club were celebrating their centenary, some quite dramatic changes were afoot at the Arms Park, with plans to convert the area into the National Rugby Stadium. The Commonwealth Games in 1958 had drawn attention to some of the problems with the Arms Park as a modern sporting venue, especially its poor drainage and

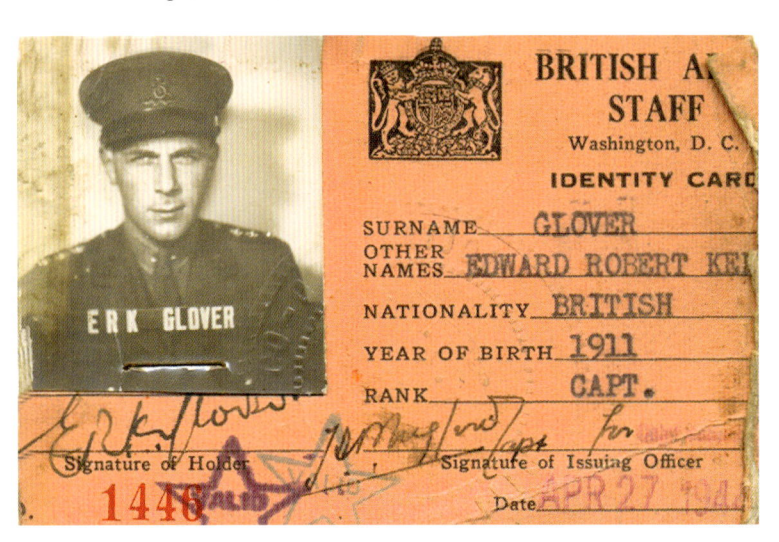

Ted Glover's wartime identity pass.

cramped facilities. The late switch of the athletics events had seen a running track being laid around the perimeter of the rugby pitch after the turf had been intensively packed and rolled so that the athletes could run on a level and firm surface.

After the Games ended, the arena had to be quickly restored to its normal state, with the greyhound track being reinstalled. In their haste, the workmen forgot to fork and break up the sub-soil, as well as in other areas of the rugby ground where the earth had been compacted. The net result was that the turf never recovered from the pounding it took and, over the next few years, which coincided with a series of wet winters, the ground became a muddy quagmire. The Taff also spilled over and covered the rugby ground with more than two feet of floodwater in December 1960, the day after South Africa had beaten Wales 3-0.

With Cardiff RFC, Cardiff Schools, the Barbarians and Wales all playing a plethora of games, as well as holding training sessions at the Arms Park, the upshot was that the ground regularly became a mudbath and an embarrassment to the Welsh Rugby Union who had long treasured the thought

Cardiff and County Club
Centenary Dinner – 1st February 1966

We, the Ordinary Members, wish to congratulate the President, Chairman, Committee and Honorary Secretary for the first class arrangements they made in respect of the above memorable function.

We also wish to Take Off Our Hats to:—

The Member from West Wales who arrived half an hour before the advertised start of the Reception armed with a bottle of milk which he proceeded to drink for the remainder of the evening

The Speaker who dealt so effectively with the first interrupter that he was heard in silence from that moment for the remainder of his speech

The Member who so effectively feigned death

The Member who thinking the above mentioned member was dead put his finger in his mouth and immediately and painfully found he was not

The Member who started to lead the Male Voice Choir through the 75 verses of Crawshay Bailey and then could not remember one of them

The Two Members who inadvertently lost their balance while sitting on the Club fender in the Hall with the result that for an appreciable time they were each 'cast' in the hearth

The Member who made every effort to cross the room to greet an old friend but unfortunately fell a yard short backwards

The Member who took the precaution of parking his car some distance from the Club and then forgot where he had parked it resulting in a long walk home

The Member who by his behaviour during the evening will henceforth be known as 'The Amorous Prawn'

The Speaker who opened his dissertation 'Gentlemen, this evening I stand...' and from that moment was inaudible above the resultant uproar

The Member who made such strenuous efforts to take the stopper out of the port decanter when passed to him until informed that the stopper had already been removed

The Member who fell down the steps of the Club into Westgate Street and then getting into his car impaled himself on the sharp end of his umbrella

A proclamation of thanks from the members to its officers.

of a stadium of their own to rival those at Twickenham and Murrayfield. Their officials, therefore, started to think about creating a National Stadium at other locations – two other sites in Cardiff were considered, as well as at an area of 90 acres at Island Farm, near Bridgend. The owners of the latter, Dunraven Estates, offered to sell their land to the Union for £19,250 so with plans for the sale progressing, the city councillors in Cardiff started to ask why the Union were thinking of taking international rugby, and the National Stadium, away from the capital city.

News that international rugby might be lost from the Welsh capital was met by a mixture of dismay and indignity by the aldermen and shop owners alike. The upshot was that an alternative plan to Island Farm was swiftly put forward, with Hubert Johnson, later to be chairman and president of

How the South Wales Echo reported the proposals.

the Athletic Club, and Ken Harris, the WRU's Treasurer, masterminding a scheme whereby the existing Arms Park would be enlarged to a 60,000 seat complex for major rugby matches, with a smaller rugby ground for club and junior matches being created on the site of the existing cricket ground. The plan would see the cricket, tennis and hockey sections of the Athletic Club move half a mile up-river to Sophia Gardens.

Soon after the revised Arms Park scheme was put forward, a few difficulties arose over the project at Island Farm with the Ministry of Transport objecting to the scheme because of potential traffic jams on the A48. With the city council fervently supporting the Arms Park redevelopment, and both the Athletic Club and Glamorgan CCC agreeing to move to Sophia Gardens, the plans for Island Farm were kicked into touch in 1964.

For Wilf Wooller and many of his friends in the Club, the news of the redevelopment of the Arms Park and the move to Sophia Gardens was very welcome. Back in the 1950s when the council was discussing the future recreational use of the Sophia Gardens Recreation Field and Pontcanna Fields, Wilf had suggested that the cricket section move to the Recreation Field where a national cricket centre could be created. The thought of no county cricket on the Arms Park displeased a few influential members of Cardiff Athletic Club and they used considerable influence to sway the council against Wooller's plans.

The thought of no international rugby in Cardiff was an entirely different matter, and the cricket lobby this time backed the Athletic Club as the green light was given to the creation of the National Stadium. On September 17th, 1966, Cardiff CC staged their last fixture on the Arms Park and, the following day, the historic wicket was ploughed up as work began on developing the new stadium for Cardiff RFC. In 1968 the freehold of the Arms Park was bought by the Welsh Rugby Union, and the National Stadium, with its new North Stand – built at a cost of more than £1 million – was formally opened on October 17th, 1970 with a match between a Welsh XV and a WRU President's XV.

In 1977 the West Stand and West Terrace, costing a further £1 million, were opened, and in 1980 the East Terrace and the Centenary Suite beneath were completed as the Union celebrated its centenary with a special match, watched by HM The Queen and H.R.H. The Duke of Edinburgh on November 29th, when an England and Wales representative team played against a combined Scotland and Ireland XV. Four years later, the final parts of the redevelopment scheme were completed – at a cost of £9 million – as the South Stand and enclosure were completed.

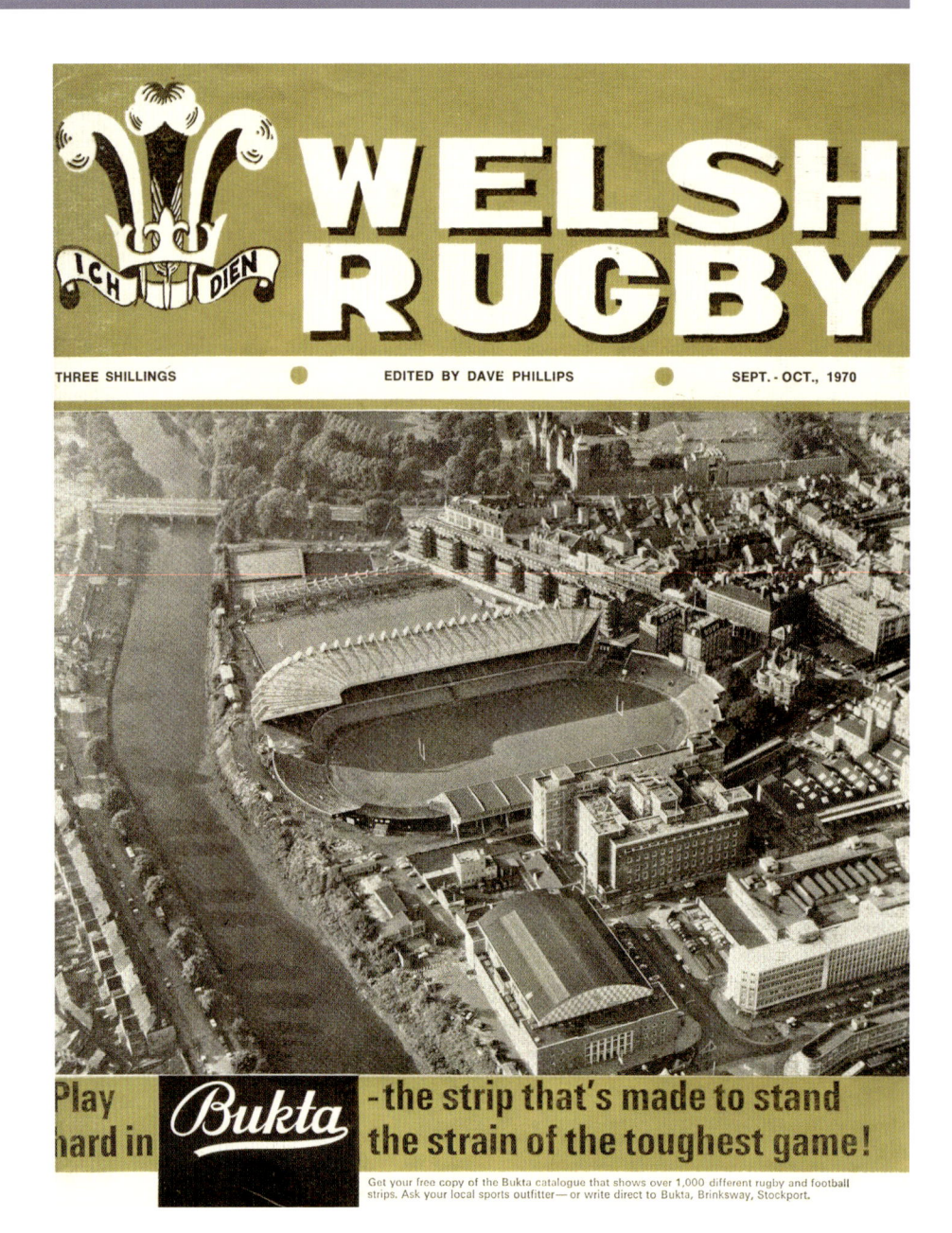

The new-look stadium complex adorned the cover of Welsh Rugby magazine and set the standard for world rugby.

It might have been business as usual during the late 1960s in the tranquil and genteel surroundings of the Cardiff and County Club but, at the Arms Park, there was plenty of change.

21

New Neighbours on Westgate Street

"The atmosphere is something that you cannot get anywhere else in and around Cardiff."

The National Stadium will probably be best remembered by sports fans as the ground where Gareth Edwards, 'King' Barry John, Gerald Davies and J.P.R. Williams won the eternal admiration of the nation, in Wales' legendary rugby side of the 1970s – a world-class team worthy of the new world-class stadium. But, as far as the members of the Cardiff and County Club were concerned, the creation of the National Stadium was a catalyst for significant changes in and around the Clubhouse, with the loss of the free view into the rugby ground, as well as the loss of the much-discussed garden and, in the minds of many, its much needed and long overdue transformation into a car park.

With the Welsh Rugby Union having taken over the freehold of the Arms Park during the early phases of its metamorphosis into the National Stadium, the committee of the Club, during October 1969, entered into a new car parking agreement with the Union for the use of the bays on either side of the access road. However, with car ownership still rising, it was felt that the number of bays would not meet demand, so other locations were considered. The Union suggested using an area under the East Terrace, one of the new stands created at the Stadium. But, after external advice had been taken by the Club's officials, this site was regarded as unsuitable because of the presence of several large tree stumps as well as an uneven surface which would cost more than £800 to smooth out.

An alternative recommendation was that almost 11 yards be removed from the bottom end of the garden and that a small cherry tree plus a couple of sections of the outer wall be removed, with a new entrance being constructed alongside a properly hard-cored area. Not everyone within the Club welcomed the loss of part of the garden, with many older members believing that it would be the thin end of the wedge and would, over time, see the

A photograph of the fire station (dated circa 1970), which stood opposite the Clubhouse on the corner of Quay Street and Westgate Street. The site is now occupied by a multi-storey car park.

disappearance entirely of the area where they had spent many hours in previous summers relaxing after a busy day at work. "I love sitting here, relaxing with a drink and looking out at the cherry trees," said Henry Lewis, with the long-serving member telling anyone, especially any younger committee members who would join him on the balcony, about the tranquil oasis in the heart of the bustling city.

But there was a huge need for more parking spaces and, with the cost of removing part of the garden being cheaper than the proposed work under the East Terrace, the green light was given to the proposals and the new facility was completed by April 1972. The work meant that the Club had a total of ten spaces in the garden as well as 35 on the access road, but even this was proving insufficient. In July 1979 the committee noted that the kerb in the garden had been removed and, on busy occasions, members were parking on the grass. It was clear that something had to be done, yet at the same time it was clear that many members no longer believed that the garden was a valuable asset. It was now overshadowed by a large concrete structure, and was no longer the setting for a convivial garden party.

The Club's President, Sir Maynard Jenour, accepted that the previous ambience had been lost and, in a circular to members, he and the committee outlined how the garden, or a site under one of the Stands, could be used for a members' car park. The plan for the garden got the green light and, in early April 1980, the committee agreed that surfacing of the garden area and an entrance to the car park would shortly follow. Many long-standing members were sad to see the garden disappear, but as the developments at the Arms Park had shown, it was time to make progress and to embrace change.

Had the Club remained in St. Mary Street, the parking issue would have been even more severe and, by using the garden, they were one of the few city centre establishments during the 1980s to have on-site parking facilities. It was a privilege which some abused, as shown by the signs warning of clamping of unauthorised cars. Some members were using it as a convenient site while shopping, and circulars were sent to members encouraging them to use the Club facilities if they were using the car park. It was a situation nicely summed up by the following entry in the suggestions book – 'Today, there were 38 cars in the car park, and I was the only member in the Club. One gentleman was eating his lunch in the car and listening to the wireless. Probably not a member.'

There was certainly much more sadness about the loss of the view into the Arms Park. For many years, the gathering in the Club on international days and the groups of spectators on the veranda and upper balconies had been most convivial, and had been a great way of bringing together members, their families and friends. They were also good money-spinners, with the well-attended lunch during March 1951, when Wales played Ireland, costing five shillings. However, the Club Manager at the time got rather carried away with things and the bar was opened before the legal licensing hour. The matter came to the attention of the President, Sir Robert Webber, who, as chairman of the Bench, reminded all concerned that the law had to be complied with and that no repetition should happen in the future.

This was duly taken on board for the match in October 1951, when Wales met South Africa, and arrangements were also made to serve coffee in the card room and for members to bring their sons in for a cold lunch and a seat to watch the game on the reading room balcony – now the bar. Duckboards were also provided on the verandas for standing room, and the roof adjoining the Secretary's flat was opened to ladies, who were also advised not to stand on the sloping roof in case it collapsed.

Over the years, a few wags had jokingly said that the only reason the old Arms Park had developed a U-shape of stands, with a lower terrace at the eastern end, rather than being completely ringed by large stands was to give the privileged members of the Cardiff and County Club an exclusive view of events in the Arms Park. The real reason that the Arms Park Company

Limited had not built up the eastern end was that they were concerned about the effect a completely enclosed structure would have on obstructing the winds which helped to dry the ground after it had been flooded by the overtopping of the Taff.

The scheme for the National Stadium, however, embraced improved drainage as well as new structures. Under the Welsh Rugby Union's grand masterplan, the old East Terrace was earmarked for demolition in the winter of 1979-80, followed by its replacement of a much larger two-tier structure with function rooms beneath. After the publication of the plans by the Union, the Club's committee met to discuss possible ways of still watching the matches from the Club's premises.

The Club initially approached the Union and enquired if the proposed new structure could be lowered, or built as a single-tier structure. Not surprisingly, this request was rejected, so the committee set about gathering alternative proposals. These included having an additional structure, up to 18 feet high, on the roof of the Club where seating could be provided for members. Concerns were immediately raised about the costs of safely securing such a structure, as well as its maintenance, and whether planning

A view of the rear of the Clubhouse showing the garden and the parking bays.

permission would be granted by the city council. Given these concerns, this idea was dropped.

Other alternatives included building a structure at a cost of £40,000 which, almost limpet-like, would abut the rear of the proposed Terrace. Another was for a free-standing structure in the Club garden with access via a staircase. The likely cost was £35,000 for a 75-seater stand and £45,000 for a 150-seater structure. In both case, the costs were deemed prohibitive so attention then switched to another much more elaborate scheme where there was an element of income-generation.

This involved a new building in the Club garden, with an access by bridge from the Clubhouse. It would be surrounded by parking spaces for 16 vehicles, and on its lower level would have two squash courts, a sauna and 17 *en-suite* bedrooms. The upper floor would also have a sloping roof to provide viewing of rugby matches for up to 400 spectators. It was a dazzling proposal which also embraced some of the parking issues, as well as providing additional recreational facilities and accommodation which could generate valuable revenue. Many other clubs, especially in London, had developed along these lines, and several committee members were quite attracted by this scheme.

However, the stumbling block was the massive expense. At the very best, the income generated each year was estimated at £26,000, but the cost of building such a structure was likely to be in the region of £350,000 to £400,000. The Club's committee were concerned about the financial implications and the fact that the scheme could leave the Club in debt for several years. As a result, the scheme was rejected and, as events have subsequently unfolded with the creation in 1999 of the Millennium Stadium, it was a very wise move as the proposed structure in the Club's garden would have become a white elephant after the construction of the completely enclosed Stadium.

The net result was that the Club decided against building any additional seating or viewing areas and the last rugby international to be viewed from the Clubhouse balconies was the game between Wales and England on March 17th, 1979. Within a few weeks scaffolding had been erected and construction work began on the new two-tier structure, thereby bringing to an end a long chapter in the history of the Club.

However, the idea for the creation of modern leisure facilities adjacent to the Clubhouse resurfaced in plans for the conversion of Jackson Hall. The city council had used the building for its Juvenile Employment Bureau until April 1967 when the offices moved to Magnet House in Kingsway. The premises were then used by the City Transport Department as offices and a canteen for bus crews following the demolition of similar premises in

Wood Street. However, in March 1973, the Transport Department vacated the premises and the building was put up for sale.

On the other side of Westgate Street, the fire station was demolished in 1973 and a multi-storey car park, accessible from Quay Street, was built in its place, and other new buildings were erected in Guildhall Place. With the Council using Jackson Hall as a set of temporary offices, as well as a polling station and an overflow for people registering for unemployment benefit, there was great doubt over the building's future, and some Club members were fearful that, if it were sold, they would be dwarfed by yet another large premises.

These fears surfaced during the autumn of 1976 when the council announced that they would be demolishing the building and using it instead as the new site for the Ebenezer Welsh Congregational Church, situated in Charles Street, which was under threat by the proposals for the creation of

Jackson Hall, as seen in the late 1970s.

St. David's Shopping Centre. Built in 1855, the church's distinctive facade had been created using ballast from coal ships returning from the Middle East, and this was one of the reasons behind the Church receiving Grade II listed status.

To the relief of Club members, the same occurred to Jackson Hall, but only after the Secretary of Wales John Morris had stepped in to veto plans for the Hall's demolition. He said: "The red bricks of Jackson Hall and those of the adjoining building of the Cardiff and County Club establish a colour scheme which is not repeated in any other part of the city streets. Jackson Hall is unique in the city."

With the future of the building secured, the City Council then sold a 99-year lease to Hamard Catering Ltd. of Barry in October 1978, so that the property could be converted into a health club called Jackson's, aimed at providing office-based businessmen and others working in the city centre with a place for exercise. To a large extent, it was a 20th century version of what many of the pioneers of the Cardiff and County Club and the Racquets and Fives Club had undertaken.

The Angel Hotel and, to the right, Westgate Street as seen in December 1972.

22

A Load of Old Rot

"It must surely be the last place in the city centre where you walk in and are warmly greeted by a glowing coal fire!"

The loss of the viewing area over the Arms Park, the conversion of the garden into a car park and the future of Jackson Hall, were only three of several issues facing the Club during the 1970s and 1980s. Indeed, a number of major financial matters reared their head, together with concerns about the structure and maintenance of the Clubhouse itself, which all combined to raise major questions about the redevelopment of the old building and viability of the site in Westgate Street itself.

Since the First World War there had been only a few changes to annual subscriptions. Subscriptions had been raised from ten guineas to 12 guineas in 1954 – the first rise since 1919 – before being increased again to 16 guineas in 1961. The reason for the further increase was that the Clubhouse was considered by the committee to have fallen into rather a shabby condition and immediate expenditure was required on a number of essential but dilapidated items.

The officials estimated that £1,600 would be required to decorate four-fifths of the Club's premises and a further £1,200 would be required for replacing the worn-out carpets in the main rooms. Funds were also needed for the provision of central heating throughout the building as well as overhauling the dumb waiter. The annual wage bill had also risen by about £2,000 since 1949 with waitresses being paid £12 per week compared with £7 2s 6d in the years immediately after the war. The overheads of lighting and heating had also risen by £400 a year.

New furniture was required; the existing furniture, together with the fittings, had been valued at just £250 on the 1961 balance sheet. The building itself had been valued at £11,000 – precisely the same amount it had cost to build back in 1891. So, with a surplus of income over expenditure in 1960 of just £61, it was clear to the committee and to President 'Tip' Williams, that the only way to fund these essential changes was through raised subscriptions.

Naturally in a Club whose members appreciated little change and preferred the *status quo*, there were a few grumbles about the rise to 16 guineas, especially as it was the second steep rise in the space of seven years. But within a few weeks, there was something far more serious for the members to be concerned with, as dry rot was discovered throughout the building, together with infestation by furniture beetles. Richardson and Starling Ltd, timber experts from Winchester, issued a five-page foolscap report detailing the problems and extent of the dry rot, from the roof timbers via the attics to the billiard rooms and down into the cellars and the coal house. The furniture in the Manager's room, together with that in the kitchen was being eaten away by the beetles, there was fungus in the urinals, and fungal spores in the garage at the rear of the building. There were also concerns about the external brickwork which was crumbling badly in places.

The increase in subscription income helped to pay the £155 for the restorative work, as recommended by Richardson and Starling. This involved saturating the building timbers with a combined fungicide and insecticide,

A view of the Clubhouse's foyer, with its welcoming coal fire, during the Chairman's Cocktail Party in December 2015.

while the brickwork was pressure-sprayed by a silicone-based water repellent solution. The fungal spores were also burnt off by a heat gun, before another fungicide spray was applied.

The poor state of the building's exterior was a much more pressing concern and Club Architect, Alex Robertson, undertook an extensive survey. The roofing slates were found to be in poor condition, whilst the walls and chimneys needed repointing. Many of the door and window surrounds were also defective, partly through wear and tear, but also because of poor construction – 'some stones having been laid with the grain on edge instead of being bedded along it' – believed at the time to probably have been caused by the labour unrest in 1891 and fewer checks being made of the builders' work. There was sundry damage to pipes, gutters and ornamental ironwork, and some of the previous attempts to make good some of the stonework had used an inferior cement mixture.

Robertson also made recommendations for a series of internal structural changes, including the removal of one of the three staircases leading to the basement, as well as additional cloakroom accommodation for those using the basement function rooms, later known as the Jubilee Suite, including additional lavatory provision for the ladies who used the room. He also noted the poor state of the wooden fence next to the parking bays and garden, and the damage caused by people scrambling over in order to gain free access to Cardiff Arms Park.

The estimated cost for all of this repair work and refurbishment was in excess of £2,000, so a Development Committee was created to deal with these matters as well as the complex issues of planning which was a result of the coming into being of the Town and Country Planning Acts of 1947 and 1954. Legal advice was sought for what had become through the new statutes a series of very complex issues. The outcome was that no application would be made for permission to carry out any of the structural changes proposed by Robertson as there were likely to be severe restrictions imposed.

There were other issues going on at this time, relating to the road layout of much of Cardiff, and the possibility of major redevelopment in the central area, including the area around Westgate Street and the site of the Clubhouse. This stemmed from the report, known as *Traffic in Towns*, which was published for the Ministry of Transport in November 1963 by a team of civil engineers, architects and town planners headed by Professor Sir Colin Buchanan. The team had been commissioned in 1960 by Ernest Marples, the Transport Minister in Harold Macmillan's

Ally Robertson.

government, to consider ways of improving the existing urban road network in the UK in a bid to relieve congestion and to mitigate other threats and possible damage caused by motor vehicles.

High Rollers

It had not been unusual to see a number of Rolls-Royces in the Club's car park but, in the case of Arthur Sansom, they could have all belonged to him.

A member of the Club from 1965 until his death in February 1997, Arthur was the founding chairman of Qualitex Printers, one of the city's leading lithographic plate and printing companies.

In his youth, Arthur had been a keen cyclist but it was four wheels rather than two which attracted his attention in adult life, with Arthur becoming a collector of fine cars. In all, he owned nine Rolls-Royces and Bentleys, some of which he proudly drove to Westgate Street when meeting up and dining with his many friends at the Club!

Traffic in Towns gave planners a set of policy blueprints, including traffic containment and segregation, which could be used in urban redevelopment schemes, as well as in plans for new roads and car-free precincts. For the city planners in Cardiff, it represented an ideal opportunity to create a series of new roads and an inner bypass, which would help to alleviate the problems caused by the sheer weight of traffic on the old roads, especially in the central areas and in the vicinity of the Castle.

Professor Buchanan was commissioned by the city council to prepare a plan for a series of roads which would ease these problems and pave the way for a central redevelopment scheme. Between 1966 and 1968 he produced a schedule of plans for a series of new roads and urban motorways, including one known as the Hook Road, which would scythe its way from the as-yet unbuilt junction 31 on the M4 and through a number of housing areas in the northern suburbs.

To an extent, the Club's Development Committee were well ahead of the game, having discussed on August 13th, 1964 how the Westgate Street site could be redeveloped. The outcome of their discussions was that the site could be offered to a development company or a commercial organisation which would look to develop the site for its own interests, with a building some seven to eight storeys high, with the Club taking the top two floors, and in the words of the committee 'with a fairly extravagant and luxurious layout to justify giving up what we have at present.' The objectives behind their discussions had

The Clubhouse, and the bus shelters!

been, in no order of importance, to make the best possible use of the site, to avoid compulsory purchase, to provide improved and modern accommodation for Club members, to share in the equity-free rent, to hold the reversionary freehold interest, and to avoid making any capital distribution.

These ideas about a new block of offices, with the Club occupying the upper two floors, were incorporated by the City Council in their Centreplan 70 Project and, on July 21st, 1970, the Club's Development Committee decided that a planning application for redevelopment of the Club's Westgate Street site was essential before a developer or the Council stepped in with a compulsory purchase order. Discussions also took place with the Welsh Rugby Union about how any redevelopment might affect the rights of way on the access road leading into the Arms Park, as well as with Council officials who had considered realigning Westgate Street.

However, there was a large outcry from the residents of Cardiff over both the Centreplan 70 scheme and the Hook Road scheme proposed by Professor Buchanan. This was not surprising given the fact that the construction of the Hook Road was likely to result in the demolition of some 1,800 houses, while the Centreplan 70 scheme was criticised from a number of quarters, especially by those who were concerned about the creation of a concrete jungle in the heart of a city with many beautiful and unique buildings. As arguments raged in City Hall, the Club gathered further information about a redevelopment of the Westgate Street site but, late in 1972, plans for the Hook Road were scrapped, and the following year, the Council also ditched the Centreplan 70 proposals. It meant that, if the Club were to redevelop their site, they would have to produce their own scheme and find a partner themselves.

It was at this stage that a scheme was put forward by architects Osborne Webb and Partners, which had the provisional involvement of the Welsh Rugby Union. It embraced several of the ideas previously discussed by the Development Committee and involved building a tail to the existing building which would be five or six storeys high and running alongside the East Terrace. The Club's dining room, bar and reading rooms would be on the top floors, all looking into the Arms Park, and the lower floors would comprise offices which would be rented out. At ground level, there would be car parking spaces, plus a private entrance for Club members, a delivery area, and a small function room for ladies. The block would also incorporate a Secretary's flat, some other bedrooms for staff, an extensive kitchen and a wine cellar.

John Webb, the Cardiff architect who proposed a redevelopment scheme for the Clubhouse during the early 1970s.

Once this had been constructed, the old Clubhouse would be demolished, with further shops and offices being created on the frontage of Westgate Street, and a wider and improved access road into the Arms Park would also be created. The Welsh Rugby Union were, at first, quite delighted by these proposals, especially for the access road and, together with the rental opportunities from the shops and offices, it represented a decent development partnership for them and the Club. Both parties agreed on October 24th, 1974 to make a joint planning application.

The sticking point, however, lay in the cost of the scheme and who was going to underwrite the building work. The costs estimated by Osborne Webb were in the region of £3 million, and lengthy discussions subsequently took place with the Union as well as with a property developer. However, a satisfactory agreement could not be reached, at which point the Union withdrew their support for the scheme. Without their backing, the grandiose project was a non-starter and the Club were left back at the drawing board, with an aged Victorian building, badly in need of repair.

A working party, under the chairmanship of John Turner, was convened to assess the situation and to consider a range of options. One possibility was to sell the property and to move elsewhere in the city centre, or even out of town, but this was rejected. Turner also prepared a report on the general condition of the fabric of the Club, in an attempt to produce a blueprint for a programme of maintenance over the next ten years. It covered the issues of the roof slates, the dry rot, the electricity and general ventilation, besides recommending that the bedrooms on the upper floor could be brought up to date and used as in other clubs.

The Turner report also found that, of the 16,000 square footage of the building, only 4,000 square feet were in regular use. Consequently, a special meeting was held on January 25th, 1980 when a lengthy discussion took place about a large portion of the Clubhouse not being used, especially the upper floors. The outcome of the discussion was that, after consultation with the Trustees, the Club should seek a tenant who could use the top two floors, with the Club concentrating its activities on the ground floor and basement. Advertisements were placed and soundings were made with several interested parties but, at the end of the day, no firm offers were forthcoming.

All concerned at the Club were increasingly frustrated by the situation, especially as the building continued to suffer problems of disrepair. In 1982 the council's fire prevention officer again demanded that certain measures were put in place, including a fire escape and further fire doors, before the licence was renewed. These cost around £40,000 and the finances took a further hit shortly afterwards when the environmental health officer reported that the balustrade at the front of the building was in danger of collapse into Westgate Street to the danger of passers-by. Other issues existed in the kitchen and laundry, where some of the equipment was almost 100 years old and in need of replacement.

At a time of an ever-worsening financial climate, this was expenditure that the Club could have done without. To a large extent, the Club had been helped during the 1960s by a raft of new members following the closure of the Exchange Club in Mount Stuart Square in 1958. Their support and subscriptions helped the Club to ride out some difficult years but, by the end of the 1970s, new and even more severe financial problems were looming. The subscriptions and Club prices were therefore raised in 1980 in a bid to generate more income. However, it prompted more resignations than new members and, as the Club's 125th anniversary approached, it looked as if there might not be much of an organisation around in 1991 to mark this event.

23

Hurrah to the Staff

"It's the people who make the place."

The 1980s were years of fundamental change to the economy not only of Cardiff but to south Wales as a whole. For several years, the coal and steel trade had been on a steady and unrelenting decline. The last coal exports from Cardiff Docks had taken place in 1964 and within a few years the areas at the southern end of Bute Street and around Mount Stuart Square were starting to resemble a ghost town. In 1978 the East Moors steelworks, once dubbed 'Dowlais by the sea', ceased operations. It had been opened in 1891 with great pomp and ceremony by the Marquis of Bute, and the industrialists of the Victorian era believed that the new complex would be in existence and provide employment for many generations. The same was said about the coal mines in the south Wales valleys but, by the 1980s after years of cut-backs and reductions in manpower, the majority of collieries across the region were also silent.

Running parallel with the decline in the coal and steel industry was the major drop in trade at Cardiff Docks and, despite a small increase in activities during the early 1980s, the Port Authority had significantly cut back its operations from the heady days when 'coal was king'. Cardiff of the 1980s still had a vibrant tertiary sector with a strong banking and financial presence, supported by many offices as well as successful shops, and popular places of entertainment and leisure. These, together with the universities and the devolved Welsh Assembly, were all subsequently at the forefront of the regeneration of Cardiff during the closing years of the 20th century.

Like other depressed dockland areas in the UK, the changes in Cardiff Bay came about only after the creation of an Urban Development Corporation as Mrs. Thatcher's Conservative government tried to grapple with the inner city crisis which had seen an upswing in violence, crime and deprivation in many, if not all, of the country's major centres of population. The Cardiff Bay

A view of the Clubhouse from the car park, taken in 1991.

Development Corporation, created in 1987, was one of several partnerships between public and private capital that stimulated the regeneration of these once thriving and bustling areas of people and commerce.

For the Cardiff and County Club, as a private gentlemen's Club, there was no lifeline from national government or generous hand-outs from the local authority. If they were going to tackle the financial problems which they faced during the 1980s, they simply had to solve the issues themselves and create a life-saving injection of cash from their own resources. For several years, the Management Committee, the Trustees and the President, Sir Maynard Jenour, had been worried by aspects of the Club's antiquated system of book-keeping and financial control, as well as the over-reliance on subscriptions to provide funds for expenditure.

At the AGM on May 24th, 1979, Sir Maynard Jenour informed members that the Club would be re-introducing a debenture scheme. As he outlined in his address, it had worked before and he hoped it would work again. But, if it did not raise the £40,000 required, the committee were prepared to sanction a direct levy. As it turned out, the debenture scheme raised just £11,450 from 63 members and, at an extraordinary general meeting in

January 1980, a levy was recommended of 100% equivalent to the annual subscription, with an increase in entrance fees for future members.

Further consultations took place with financial advisers and what subsequently unfolded at the AGM on June 23rd, 1983 was the introduction of a mandatory interest-free loan of £300 payable by instalments. There was some disquiet and opposition to the proposals and, by January 1984, the Secretary had received two dozen resignations, and 14 members, though paying their £153 subscription, had not responded to the demand for the first instalments of the interest-free loan.

Nevertheless, the mandatory loan arrangement brought financial stability and allowed the Club to undertake several tangible improvements to the décor, as well as a complete rewiring of the premises, plus the installation of both central heating and a security alarm. The basement was refurbished and other improvements occurred in the bar, reading room and card room.

There were improvements as well to financial housekeeping, with a much improved system of financial control being introduced, with computers producing accurate monthly accounts, as well as helping the Manager to keep a close eye on the income from the dining room and bar. By 1986 the Club was able to report a deficit of £32,404 but this was only after £44,276 had been spent on repairs and refurbishments. The Club had £122,303 from the mandatory loans, plus £12,637 through gifts, £9,292 from entrance fees, and a further £91,242 from subscriptions. The deficit continued to be reduced in subsequent years and, in more years than not, a trading surplus was reported.

The improving financial position of the Club during the 1980s mirrored many of the other changes to the economy of south Wales during the tenure of Margaret Thatcher as Prime Minister between 1979 and 1990. The policies of her government included, among other things, the devolution of government decision-making to local authorities, the privatisation of certain national industries, in addition to reducing the influence and power of the trade unions. These policies changed many aspects of the fundamental geography of the south Wales coalfield and many of the traditional Labour strongholds, especially in the valleys, vehemently opposed what the Conservative government attempted to do. It was, though, part of a modernisation trend as British-based industries attempted to become more competitive within the global economy, while the injection of fresh capital and inward investment by multinational companies, as encouraged by the Welsh Development Agency as well as the Cardiff Bay Development Corporation, saw the emergence of hi-technology industries and a move away from reliance on heavy industries and port-based activities.

The 1980s therefore saw the seeds being sown of new initiatives and businesses which subsequently flourished as south Wales moved into the 21st century. But, as well as these seismic shifts in the region's economy, the decade nearly saw tragedy with 'The Iron Lady' being very fortunate to escape an assassination attempt by the Irish Republican Army during the early hours of October 12th, 1984, when the Conservative Party were holding their annual conference in Brighton. The Prime Minister and other leading Conservatives were all staying at The Grand Hotel, where a few weeks earlier a bomb had been planted in the bathroom of the suite in which Mrs. Thatcher was scheduled to stay. It detonated at 2.54am on that fateful day, with the Prime Minister – who was still awake at the time and preparing her speech for later in the day – escaping from the rubble with her husband Denis. Tragically five people were killed, and a further 31 were injured.

Among the death toll was Sir Anthony Berry, a member of the Cardiff and County Club who, from 1955, had been managing director of the Western Mail and Echo Ltd. before becoming a politician and reaching the upper echelons of the Conservative Party. Educated at Eton and Oxford, Sir Anthony was the sixth and youngest son of Merthyr-born newspaper baron Lord Kemsley whose vast publishing empire included the Cardiff titles. After serving as a Lieutenant with the Welsh Guards, he began his training in the Press world, initially with the *Daily Despatch* in Manchester, before moving to Cardiff. On successive days in October 1952, his name appears in the Club's Visitors' Book as he dined initially with Sir Robert Webber, before the following day enjoying the company of Frank Webber, the vice-chairman and general manager of Western Mail and Echo Ltd.

On the retirement of Sir Robert in 1955, Anthony became the managing director and, together with his first wife, moved to Miskin Manor and then later to Llantrithyd House near Cowbridge. Through his marriage to the daughter of the Fourth Baron Fermoy he also subsequently became an uncle of Diana Spencer, later the Princess of Wales. During his time at the helm of the *Western Mail*, Anthony was a regular visitor to the Club and, in 1962, he was appointed High Sheriff of Glamorgan. Two years later, Roy Thomson acquired Kemsley's newspapers and this prompted a change in Anthony's career, plus a move away from south Wales. In 1964 he was elected MP for Enfield, Southgate, before rising to influential and senior positions within the Conservative Party, and being knighted in December 1983. At the time of the party's conference in Brighton in 1984, he was a whip in Mrs. Thatcher's government but his life was sadly cut short by the bomb on the final morning of the conference.

His tragic death was deeply felt by many senior members of the Club. Among these was Alex Robertson, a member since 1947 and known to all as Ally, having fulfilled, at different times, the positions of Chairman of the committee, Honorary Architect and Honorary Archivist before taking over as Secretary in 1981. Ally was in post when a rare case of poor behaviour took place in the dining room. It involved a senior figure in the legal world who, while dining, became embroiled in a quite contentious conversation with a fellow member, before standing up and saying "Don't be such a bloody idiot!" and pouring a jug of water over his dining companion. He was duly suspended for three months from the Club premises.

Derek Tantum and his staff colleagues.

Ally Robertson's appointment as Secretary had followed some difficult years during the 1970s when a number of other people filled the position. One was in post for only six weeks after the committee found that there had been numerous and expensive phone-calls being made on the Club's telephone to overseas numbers. Another eminently more suitable person appeared to be Captain Eastwood, who joined the Club in August 1978 after his release from his Army unit but a year later he departed after securing an appointment with the Territorial Army Association of Wales. His sudden

A light-hearted take, by cartoonist Austin Thomas, on a famous episode at the Club.

departure meant that, for a while, the committee had to assist in the day-to-day running of the Club, but much needed stability came about with Robertson's appointment and he duly served in this capacity until his death in 1991.

Derek Tantum was another stalwart of the Club through the difficult years of the 1980s and a man who helped spark the revival into the next decade and beyond. Appointed Catering Manager in 1981, he was promoted to Club Manager the following year and subsequently drew on all of his experience working in hotels, restaurants and nightclubs. The son of a Porthcawl butcher, he had emigrated in 1965 to Australia but subsequently returned, and the Club was eternally gratefully that Derek accepted the offer, from Chairman John Turner, to work in the Clubhouse. In the ensuing years, Derek, together with Head Chef Mark Jones, significantly raised the standard of cuisine and service.

'Mr. T', as he was known to all and sundry, ran the Club with a rod of iron and strictly enforced the dress code. Even on the coldest of January days, he would ensure that nobody dared walk into the bar still wearing an overcoat, or entered the dining room carrying a pint or wine glass. However, he could be persuaded by a pretty face to turn a blind eye as, on one international weekend, when he allowed the five pretty daughters of Adrian Lloyd-Edwards to stroll into the dining room with large gins and tonic in cups and saucers!

Edna

The longest-serving member of staff during the post-war era was Edna Cross. She joined as a waitress in 1946 and retired in December 1980 after many years as head barmaid. A special testimonial fund was raised on her retirement. In his letter to members, the Chairman, Michael Gibbon reflected how 'after thirty-five years of service to the Club, Edna is retiring. There can be no member of the Club who has not the greatest respect and regard for her. She has given excellent service – efficient, courteous, discreet and uncomplaining at all times.'

Edna, Anne and Geevers.

Edna, who lived on her own in Grangetown, was a trusted *confidante* for many members and with a twinkle in her eye, she would gleefully field any telephone calls from irate wives seeking information about the whereabouts of their husbands who had called into the Club and were late in returning home. "I can't see him at all. I don't think he's here anymore," was Edna's standard reply, often delivered when looking at the errant husband straight in the eye on the other side of the bar and quietly passing him another drink!

A special presentation to Edna took place by the President, Sir Maynard Jenour, at lunchtime on December 19th, 1980 at which she briefly reflected on her happy years at the Club, as well as on the staff annual outings which had been held since the 1950s. One such outing was on Whit Sunday 1959, when 40 members of staff and their spouses travelled to the Hotel Majestic in Cheltenham and they enjoyed a roast chicken lunch before strolling around the elegant Regency town.

Another of Austin Thomas' humorous cartoons.

Mr. Tantum's high standards have been maintained by a number of others, notably Sally Johnson, a very efficient dining room Manageress, and in recent years Dawn Said, who has led the waiting team which has, on occasions, included a number of members of her family. Angela Flatt and Mary Saleh were popular waitresses, with Mary overseeing the wines for many years and, behind the bar, Joan, Sandy and Barbara were friendly faces for members, with, more recently, Peter Wilcox cheerfully overseeing operations.

In the kitchen, Lee Grubb and the current Head Chef, Lee Thomas, have continued the traditions set by Mark Jones. Lee Grubb joined the Club shortly after leaving school, and was Sous Chef to Mark for over a decade. On becoming Head Chef, he expanded the menus besides adding his own inventive touches. In the past three years Lee Thomas, ably assisted by Camilo Estevez and more recently by Dan Howell, has taken the cuisine to what many members believe to be the finest level it has been in the Club's history. In addition, Lee and his team have worked effectively alongside several award-winning guest chefs at special evening functions, with the guests complimenting Lee on his culinary skills and efficient management of the kitchen operation.

Turning to the clerical staff, Mair Forsdyke, as Club Administrator, was, in the words of a long-standing member, "a lovely lady who was the backbone of the Club for a number of years. Beneath her charming and pleasant exterior, lay a spine of steel, and she helped guide the Club through some difficult times." After Angela Williams had a spell as General Manager, Gill Thomas has latterly continued Mair's good work, with Gill's friendly and helpful manner endearing her to colleagues and members alike, besides being very

Derek Tantum (right) with staff colleagues: Alison Simpson, Joan Holdstock, Maureen Lewis, Sally Johnson, Mary Saleh, Angela Flatt and Blanche Montes.

Peter Wilcox (left) hands out drinks, at the Chairman's cocktail evening in December 2015, to (from left to right): Maggie Newton, Judy Watson James and Simon Hooper.

willing to turn her hand to helping out behind the bar or serving in the dining room.

Gerry and Charles were two long-serving porters, with Gerry in particular showing good humour in performing his duties which, by their very nature, could be quite mundane. He also showed great wit, and a certain amount of discretion one evening when, during a function an amorous couple ventured upstairs and were subsequently discovered by Gerry *in flagrante delicto* on top of the snooker table!

Back in the days when some staff lived in, occupying the flats on the upper floors, there was a tradition for a staff outing with Whit Sunday, 1966, seeing staff from the Club and some of their families visiting Tenby, with lunch and high tea at the Clarence House Hotel. In June 1968 a group

The kitchen team in full flow, as Lee Thomas, Dan Howell and Camilo Estevez prepare the fine fare at the Burns Nicht event in February 2016.

visited the Mendip Hotel in Blagdon before taking a bracing walk along the Mendips. Roast chicken was on the menu again in June 1975 when staff and their spouses from the Club visited the Greenhouse Café in Lynton before an excursion on a Western Welsh 49-seater coach through the Valley of the Rocks. In February 1975, rail was the favoured means of transport as staff travelled to the London Palladium to watch Tommy Steele in a matinee performance of a West End show. The happy hours the Club's staff spent at these various resorts, as well as in London, were a suitable reward for their unstinting efforts at the Clubhouse.

Following the raising of the mandatory interest-free loans from members in the 1980s, the Club was able to look forward to further successful activity during the 1990s and beyond, as the capital city of Wales, under the guidance of the city council, the Government and the Cardiff Bay Development Corporation, successfully switched its local economy from manufacturing and port-based activities to those in the service sector, the arts and the world of culture.

A packed dining room at the Founders' Lunch held in January 2016.

Indeed, there were many smiles on the faces of members of the Club as they celebrated the 125th anniversary in 1991, as well as partaking in other grand functions, including a reception and cocktail party on November 28th, 1990 attended by a record number of 280 members and guests, plus a private viewing, at the National Museum of Wales, on December 11th, 1990 of the royal collection of paintings from Windsor Castle which saw the Secretary accept 493 invitations.

Those attending these gala occasions, as well as the committee, Trustees and President, were all very grateful that the Club had been blessed by having the services of many long-serving and loyal administrators who were all prepared to maintain the highest of standards as well as doing their best to get the Club out of the financial trough. As one long-standing member commented, "it's the people who make the place."

24
125 'Not Out'

"It's like walking into a first-class restaurant where you know everyone."

Given the historic link with the Bute Estate, it was quite fitting that the Sixth Marquess of Bute was the Club's guest of honour on May 24th, 1991, as the members celebrated the organisation's 125th anniversary. A lavish black-tie dinner, costing £50 a head, was held with 150 members

The Marquess of Bute (third left) is joined by (left to right): Sir Cennydd Traherne, Sir Maynard Jenour, Norman Lloyd-Edwards, Keith Mainstone and Lord Mayor Cllr. Jeffrey Sainsbury at the 125th Anniversary Dinner held at the Clubhouse in 1991.

Above and below: Club members at the 125th Anniversary Dinner.

dining across three rooms at the Club with Norman Lloyd-Edwards, the Lord Lieutenant of South Glamorgan, making the response, and the President and Chairman also saying a few well-chosen words on this august occasion at which Jeff Sainsbury, a member and the City's Lord Mayor, was also in attendance.

The grand event was held just three months after the death of long-serving Secretary Alex Robertson, known to all as Ally, who died at the age of 81 after almost 20 years of service, in various guises, as an administrator for the Club. In his professional career as an architect, he had been a founding partner of the Robertson Francis Partnership, one of Wales' oldest practices after joining his father-in-law Morgan Willmott during the early 1930s. He joined the Club in 1947 and became a familiar face in and around the Club, especially during the 1970s when his advice on a range of matters relating to the structure and up-keep of the Clubhouse was greatly valued by the management. As Donald Box, the Club's Chairman stated in the Annual Report for 1990-91: 'Ally was a great character, endearingly irascible, but a great source of inspiration and his devotion to the Club and its members was second to none.'

Like Geevers, Ally lived in one of the flats on the upper floors of the Club, and he was one of half a dozen regulars who, each evening, sat around the log-fire in the lounge, chatting about a range of topical matters, sharing a few yarns and, on Sunday evenings, gleefully tucking into large bowls of chips which Edna Cross would bring in for the bachelor party. His regular companion was George, a well-behaved dachshund who, in a rather bizarre way, became an honorary member of the Club.

It happened one damp winter's evening when Ally had called a special meeting to discuss a modest rise in the annual subscriptions. Whether it was the fact that there was little opposition to the proposal, or the weather itself, but only 11 members had assembled in the Club by the time the meeting was scheduled to start. Aware that the meeting needed a quorum of 12, Ally, with a glint in his eye, spoke briefly with a senior officer before announcing to the group that he was proposing that his four-legged friend be appointed an honorary member. His proposal was passed *nem com* before a show of hands, and paws, saw the resolution adopted!

Frank Gaskell, a Cardiff-based solicitor, was another of the regulars each evening, arriving like clockwork around 7.45pm, for either two or three whiskies which he always bought himself, steadfastly refusing anyone's generosity, each time replying "My dear boy, that's very kind, but no thanks." Around 9.30pm Frank would then return to his home in the adjoining Westgate Street flats, content that he had put the world to rights.

Ally was just one of several notable figures in the Club to pass away during the 1990s. Club President Sir Maynard Jenour died on September 1st, 1992 at the age of 87. The Old Etonian had been a prominent industrialist and public figure in south Wales for many years, besides being a member of the Club since 1933. His mother had been a member of the Beynon family, who owned several collieries and were very active shipping agents.

After leaving military service during the Second World War, he became a director, and subsequently chairman of the Aberthaw and Bristol Channel Portland Cement Company from 1946 until 1983 when the firm was taken over by the Blue Circle Group.

During the Second World War, he had joined the Royal Artillery, serving as second-in-command of a heavy gun regiment in both the UK and Egypt. After returning from the Middle East, he became High Sheriff of Monmouthshire from 1951 before, in 1974, becoming the Deputy Lord Lieutenant of the new county of Gwent. From his home in Chepstow he was, at various times, president of Cardiff Chamber of Commerce, and the Boys' Clubs Association of Wales, besides serving as a governor of the National Museum of Wales and of Christ College, Brecon and as the treasurer of the Conservative Party in Wales. Sir Maynard was knighted by Harold Macmillan in 1959.

While at Eton he had been a decent boxer and hockey player and, during the late 1920s, had played Minor County cricket for Monmouthshire, with his debut coming against Dorset in 1928 at his home ground in Chepstow. On four occasions, he was invited to stand for Parliament, but each time he turned down the approaches saying "I wish to spend more of my time, not less of it, with my family!"

January 1995 saw the death of Sir Cennydd Traherne, a proud member of the Club who, in 1970, had become the first Welshman to be made a senior Knight of the Garter. A measure of his standing and very wide circle of friends was that a service of thanksgiving in his memory in Llandaff Cathedral was attended by more than 1,000 people. An unassuming man of great charm, his business and public life after the Second World War had seen him serve, at various times, as president of Cardiff Business Club and Cardiff Chamber of Commerce, besides holding positions on the board of Cardiff Building Society, the Welsh Gas Board, the National Provident Institute for Mutual Life Assurance, and being heavily involved with the Church in Wales and the Glamorgan Hunt.

Knighted in 1964, he had also been appointed an Honorary Freeman of Cardiff and, as his obituary in *The Times* noted, he was 'a proud champion of the interests of the Principality.' David Mansel Jones, who was then serving as Chairman, eloquently summed up the feelings of all of Sir Cennydd's friends and fellow Club members when writing to his family, saying 'his record of service to the Crown and the community is unsurpassed, and all the more so when carried out with such dignity and quiet modesty. We are greatly privileged to have had him as a member and to have known him as a friend.'

Cennydd Traherne (second left) riding with other leading members of the Glamorgan Hunt (from left to right): David Crichton-Stuart, the Marquess of Bute, Evan David and Bob (R.H.) Williams.

Following the death of Sir Maynard Jenour, the Presidency was taken over first by John Cory and subsequently Glynne Clay – two men whose families had been synonymous with the growth of Cardiff Docks during the second half of the 19th century. In John Cory's case, his family were also intimately connected with the creation of Barry Docks. Born in June 1928 at St. Brides-super-Ely, John lost his father ten years later and, together with his brother Christopher, was raised by his mother. John attended Eton College, although he often tried to evade returning to the famous school by hiding on the platform when changing trains at Waterloo. Once he put eggs into his trouser pockets and then squashed them on getting into the car when he and his family were about to set off for the railway station. But he soon came to relish his time at Eton where his interests in country pursuits were kindled by his housemaster Claud Beazley-Robinson who, besides driving an ancient Rolls-Royce, hunted regularly with the Fernie Hunt.

After leaving Eton, John went up to Trinity College, Cambridge, but he did not stay the course. With his asthma barring him from National Service, he did a stint in accountancy before returning to south Wales and joined John Cory and Sons, for whom he served as non-executive director from 1949 to 1991. He became a well-known figure in the worlds of hunting, point-to-point racing and horse-breeding, as well as a regular face at the Cardiff and County Club.

From a young age he had helped Captain Homfray, Master of the Glamorgan Hunt, at the kennels. During the Second World War, the teenager acted as whipper-in for the Hunt. He became joint master of the Glamorgan Hunt during the 1960s and later served as a steward at the Bath and West Show in Shepton Mallet, as well as at Ludlow Racecourse. In 1959, at the age of 31, he became High Sheriff for Glamorgan, before becoming a Deputy Lieutenant in 1968 and subsequently Vice-Lieutenant of South Glamorgan in 1990. For most of his adult life, John was also vicar's warden of St. Brides' Church and, from 1957 until 1974, a member of the Governing Body of the Church in Wales.

For many years, he was also involved with the Vale of Glamorgan Agricultural Society and was president of the National Light Horse Breeding Society and vice-chairman of the Sport Horse Breeding Society of Great Britain. John's wife Sarah was the sister of Richard Meade, who won a trio of Olympic gold medals and, at the Mexico Games in 1968, became the first British rider to win an individual gold.

Glynne Clay was the youngest of Johnnie Clay's three sons and shared his father's love of horse racing, serving as Chairman of Chepstow Racecourse from 1982 until 2000 during which time he established the Welsh Grand National on a firm financial foundation and saw membership of the racecourse increase. He was born in 1931 at Penlline Castle, the home of his mother Gwenllïan, who was the daughter of Club President, Colonel Homfray. He followed in his father's footsteps by attending Winchester College before completing his national service with the Royal Artillery. After demobilisation, he joined his uncle's stockbroking business, which later became part of Brewin Dolphin.

For many years Glynne, who lived at St. Hilary, was a senior partner at stockbrokers Lyddon & Co, before in 1998 being appointed managing director of Rothschilds in Cardiff. He was also president of Cardiff Business Club and during his time in office, he presided over the first visit made to the Business Club of a serving Prime Minister, when Tony Blair came to the Welsh capital city.

Horses for courses

Glynne Clay is the only member of the Cardiff and County Club to have owned and trained runners in the Grand National Steeplechase as well as owning a winner on Gold Cup Day at the Cheltenham Festival. In his youth, Glynne followed his father by riding in point-to-points and being an active huntsman. After success in local point-to-points, Glynne took out a permit to train horses, largely for his father and his circle of friends. During the 1960s and 1970s several of the Clay family's horses achieved much success always carrying the distinctive silks which represented their sporting passions – red for Wales, with a pair of hoops in the yellow and blue of Glamorgan CCC.

One of their successful horses was a steeplechaser called Claymore, named in honour of the family's successful shipbrokers at Cardiff Docks. Claymore ran in the 1964 Grand National at Aintree, finishing 13th in the 33-runner race, and was ridden at Aintree – and in many of his other races – by Glynne's good friend and fellow Club member, Colin Hughes-Davies. A member of the Club from 1952, Colin later became a well-known trainer in his own right and, from his stables at St. Arvan's, near Chepstow, he became famous for his association with champion hurdler, Persian War.

Described as 'the greatest hurdler we have seen', Persian War won the premier race at Cheltenham on three successive occasions. After a moderate career on the flat with Dick Hern, Persian War first enjoyed success over hurdles with Tom Masson, before being bought by Henry Alper, a successful loss assessor but a rather overbearing owner. After watching his new purchase win the Triumph Hurdle at Cheltenham, Alper sent the horse to France, but the owner again fell out with the trainer, and Colin was asked to take over.

After travelling to Chantilly, Colin was shocked by what he discovered. "A madman, wearing breeches with huge checks, had been in charge of him," he later recalled. "Persian War was in this vast box, freezing cold. He was a sick

Glynne Clay.

horse." Back in Gwent, Colin nursed the horse back to health and, ridden by Jimmy Uttley, Persian War won the Champion Hurdle in 1968, 1969 and 1970. However, by the time of his latter victory, Persian War had switched stables again and was trained by Arthur Pitt. The disagreement with Alper had been desperately hurtful for Colin and his wife Helena, especially as they were immensely fond of this good-natured and top-class horse.

One of Glynne and Johnnie Clay's favourite horses was Sixer, a fine steeplechaser who, like several of the family's horses was named after a cricketing phrase. Sixer was runner-up to Deblin's Green in the 1973 Welsh Grand National, before the following year appearing in the Grand National at Aintree. It was a huge thrill for Glynne to again enter the owners' and trainers' enclosure at the famous Liverpool racecourse, before saddling up Sixer and giving the leg-up to Welsh jockey Taffy Salaman. Priced at 66-1, Sixer was one of 42 runners in the four-and-a-half mile contest, but he got only as far as the fourth of the 30 fences, as he was brought down by another horse, leaving Glynne and his party to watch as Red Rum recorded the second of his three triumphs in the world's greatest steeplechase.

Glynne's biggest success as an owner came on Gold Cup day at Cheltenham in March 1989 when his Observer Corps won the Cathcart Challenge Cup, which was the final steeplechase of what was then a three-day festival of National Hunt racing. Despite a starting price of 66-1, it proved to be a shrewd decision to run the horse in the two-mile, five-furlong race, open to novices and second-season steeplechasers. Observer Corps – ridden by Irish jockey Tom Morgan and trained at Ross-on-Wye by John Edwards – jumped well throughout the race before galloping up the hill to the finishing line to win the race. A measure of Observer Corps' achievement was that in second place was Norton's Coin which, the following year to the day, won the Gold Cup at odds of 100-1 for Carmarthenshire farmer Sirrell Griffiths.

With Commander John Payn firmly ensconced as an efficient and hard-working Honorary Secretary, the early 1990s saw a continuation of the improvement work to the Club. During the late summer of 1992, the refurbishment began of the lower front room into an office to be used by the Commander, as well as Derek Tantum and Mair Forsdyke, the Assistant Secretary. Mair had been working at the Club since 1986, and during Ally Robertson's illness had stepped in and kept the office running smoothly until the Commander was appointed.

During the early 1990s, the Jubilee Suite was also upgraded and, in January 1991, a widescreen television was purchased and installed in the reading room. From September that year Arthur Weston Evans was appointed as Club Archivist and, over the course of the next few years, he wrote a fine history of the first 125 years of the Club, besides setting up a proper archives room on the upper floor, rather than letting the paperwork build up and gather dust and mould while lying in boxes on the cellar floor.

There were some innovations as well in and around the Club, including the introduction in March 1991 of the St. David's Day Dinner, something which has subsequently become a fixture in the Club's calendar of events. The introduction of bus lanes, though, in Westgate Street in November 1994 met with less satisfaction, with car-driving members having a few near misses as they turned into the street from the access road, and one unfortunate member drove into a corporation bus – something that prompted the Secretary to post a missive on the notice board asking all members to take great care when driving away from the Club.

It was not, though, all onward and upward – or in the case of the motorists, carefully to the left – as there was a drop in dining room and bar takings during the early 1990s, with the committee minutes for July 1994 noting a fall of £14,000 compared with the same time period in the previous financial year. It prompted Derek Tantum, after consultation with the Head Chef, to consider

Arthur Weston Evans, the author of the Club's 125th anniversary book, with his wife Isobel at their Diamond Wedding Lunch held at the Clubhouse.

changes to the menu, with a roast each day as well as a simpler and lighter offering at a reduced price for those members who did not want to have a full-blown meal at lunchtime.

The drop in income also led to the committee recommending an additional ballot of new members in the closing months of 1994 as the overall deficit for the year was, in the absence of any fresh blood, estimated to be in the region of £20,000. Fees from new members was especially valuable income at a time when there was an ongoing programme of maintenance, as well as work being required to repair water damage to the ceiling of the snooker room, to re-point some of the balconies and parapets, to refurbish some of the urinals and to install an acoustic hood and telephone booth near the hall, as well as remedial action to counter subsidence at the rear of the building.

There was also a need to improve the close-circuit TV and security fence around the perimeter of the building following instances of graffiti being scrawled on the Clubhouse's stonework and a number of incidents of damage to, as well as the attempted theft of, vehicles parked in the Club's car park. In July 1993 the committee discussed a number of issues following reported incidents, as well as bemoaning the fact that it was well-nigh impossible to keep the gates locked when members were passing in and out so frequently.

The situation, though, did not immediately improve and, in November, the committee agreed to a car park attendant being in place from 12 noon until 4pm, and members were urged to report any unauthorised persons seen loitering in or around the car park. Sadly, these measures were to no immediate avail as in December 1993, the Club officials noted that a car had been broken into over the course of a weekend, and the following January a car was almost stolen but for the quick-thinking actions of Club members who almost caught the would-be thief.

25

The Devolution Debate

"The Club is practically my second home, and it's a place I would dearly love to visit every day."

Pride in the country's achievements may be one thing but, despite the presence in the membership of a number of supporters of Plaid Cymru, the Cardiff and County Club has never been a hotbed of Welsh nationalism, except on international rugby days!

Nevertheless, there was plenty of debate within the bar and other rooms in the Clubhouse about the vote, on September 18th, 1997, over devolution and the creation of a Welsh Assembly. It was not the first time that there had been a vote – the first, on March 1st, 1979, had seen the proposal soundly rejected by four to one, but the topic was back on the agenda from May 1997 following Labour's return to government for the first time since 1979. Labour's manifesto had included a commitment to a referendum and when it took place, 50.3% of those voting supported the idea giving the proposal the narrow majority of 6,721.

September 18th, 1997 was therefore another landmark in the history of the city of Cardiff, despite the capital voting against the establishment of the Assembly. Being the home to the Welsh Government has duly given Cardiff a meaningful and tangible focus for its status as a capital city, rather than being a capital city in name only. The brand new buildings in the regenerated Cardiff Bay have added a fresh, vibrant and very important element to the gentrification of the area which had once been the very heartbeat of the coal metropolis but had experienced a downswing since the Second World War.

It was no surprise that the morning after the historic vote, politicians were describing it as 'a very good morning in Wales' and, two days after the vote, the nation's professional cricketers, under the captaincy of Club member Matthew Maynard, did their bit for 'Cool Cymru' by winning the County Championship title for the third time in their history, and the first since 1969. Wilf Wooller, the *éminence grise* of Welsh cricket, was sadly not alive to see either of these new chapters in Welsh sporting or political history as he died after a short illness in March 1997. Indeed, he was one of several prominent people in public life in Wales, and well-known faces at the Cardiff and County Club, to pass away during this totemic year.

The first was William Crawshay, a member since 1954 and one of the Club's Trustees since 1972, who died on January 24th. Born in 1920, Sir William Robert Crawshay was a member of the family whose massive wealth derived from iron-making in the Merthyr area during the 19th century. He had served with distinction with the Royal Welch Fusiliers in the Second World War, during the latter part of which he was deployed in the Special Operations Executive and, together with two other officers, was parachuted into France in the early hours of D-Day, and into German-held territory from which he operated as an *agent saboteur* with the French Resistance.

The 24 year-old and his two colleagues had initially been set to head towards central France in the early hours of June 5th, but the rough weather in the English Channel and the heavy storms led to the brief postponement of the Normandy landings as the massive flotilla of vessels returned to their moorings in Southern England without attracting German attention. The plane carrying the three secret agents also returned to base, before setting off again the following morning as Operation Overlord commenced on June 6th.

Under the pseudonym of Major Crown, Crawshay landed by parachute close to the hamlet of Montcousinat in Chitray. From the Domaine de Montgenoux, a large stone building with links to a nearby Cistercian Abbey, he and three members of the French Resistance formed a unit known as the Jedburgh High Team. For the next six months, they undertook manoeuvres behind German lines, including sending back details of enemy positions using a radio which Crawshay had smuggled in his backpack when landing on French soil. The unit also helped other members of the Resistance to carry out sabotage work using weapons which were dropped by parachute from Allied planes, as well as aiding the swift and safe return to England of pilots who had landed behind German lines.

Crawshay's brave and fearless actions, together with those of the rest of the Jedburgh High Team and the other units in the Special Operations Executive, played a key role in the defeat of the German troops in the Indre *department* in Central France and, on June 21st, 1945, Crawshay – who by now was a major – was awarded the Distinguished Service Order for his gallantry. He subsequently returned to civilian life in south Wales and held various positions in public life, including being a board member of the National American Welsh Foundation, as well as serving on the boards of the National Museum of Wales, the Welsh Arts Council, and the Council of the University of Wales.

> ### Henry Lewis
>
> Henry Lewis, a stalwart member of the Club, died in February 1997 at the age of 63. Henry, who had been a member of the Club since 1958, was a member of the family who owned the Taff Wagon Engineering Company and other businesses manufacturing railway tracks and double-decker buses. His family had a long and proud association with the Club, with his grandfather having also been one of the people who helped to pay for Captain Scott's attempts in 1910 to reach the North Pole. Along with other supporters and sponsors Lewis' grandfather dined with Scott in the Royal Hotel the evening before he set sail from Cardiff on the *Terra Nova*.
>
> Henry was educated at Shrewsbury School before reading law at Trinity College, Oxford, and later rising to the rank of Squadron Leader in the Royal Air Force. He was a most enthusiastic cricketer, playing for among others I Zingari and the South Wales Hunts. It was Henry who conceived the idea for The Cricketer Cup, a competition for the old boys' teams of leading public schools – an idea he apparently came up with while drinking dry sherry with friends at 5 o'clock one morning!
>
> Indeed, Henry was a noted *bon viveur* and, as a member of several clubs in London he also mixed with many figures in the capital's social and sporting scene. Through his cricketing connections, he became friends with Denis Compton, the Middlesex and England batsman, and while dining with 'The Brylcreem Boy' at Bentley's in Swallow Street, Henry found a black pearl in one of his oysters. He immediately gave it to a waitress who subsequently had it made into a ring and always remembered to thank Henry for his kindness when he dined with other friends in the fashionable restaurant.
>
> He served as High Sheriff of Glamorgan in 1958, and was chairman of the Boys' Clubs of Wales from 1979 until 1989. Despite his commitments in London, for many years Henry lived at 'Cliffside' in Penarth and split his time between south-east England and south Wales, regularly travelling by train to and from Cardiff. He did, though, have mixed feelings about returning on the late afternoon services when high speed trains were first introduced. "It's turned a decent five-gin journey into a three-gin one," he would say when asked how his journey was by his wide circle of friends on arriving in the bar at the Cardiff and County Club.

'JBG' (third left) receives a special presentation from officials of Glamorgan County Cricket Club, including: Judge Rowe Harding, Wilf Wooller, Ossie Wheatley, Gwyn Craven, Viv Jenkins and Phil Clift.

April 1997 also saw the passing of Bryn Thomas, or 'J.B.G.' as he was known to many, the award-winning rugby journalist who for 36 years was chief rugby writer of the *Western Mail*. From the 1950s and into the 1980s, Bryn spent many months abroad covering Wales or the British Lions on their overseas tours, He was Sports Editor from 1965, and assistant editor during the late 1970s.

Bryn also found himself on the front page of the newspaper for several days in November 1978 after the controversial victory by the All Blacks over Wales in Cardiff during the New Zealanders' Grand Slam tour of the British Isles. His astute eye, in those days before a plethora of multi-angled and slow-motion television replays, saw the All Blacks' locks, Frank Oliver and Andy Haden, deliberately dive out of a line-out and gain their side a match-winning penalty. Having witnessed, as a young man, Wales' victory over New Zealand in 1935, and having reported on Cardiff's famous win over the 1953 All Blacks, Bryn thought he was about to see another great chapter in Wales' rugby history, but the subsequent penalty kick from Brian McKechnie saw New Zealand snatch a narrow and controversial victory.

Enraged by what he had witnessed from the seats in the Press Box at the Arms Park, Bryn duly accused the All Blacks of cheating, with his words on the front page of the *Western Mail* being flashed around the world. Many thought that it was a case of sour grapes, and that he had gone over the top, but Bryn's words were later borne out when Andy Haden wrote in his autobiography about the plan to dupe the English referee, Roger Quittendon.

Born in Pontypridd, Bryn had worked initially as a civil servant at Cardiff City Hall before sending match reports and other features to local newspapers. He joined the Royal Navy during the Second World War and, after getting married in 1941, rose to the rank of First Lieutenant on a minesweeper before returning to south Wales and joining the staff of the *Western Mail* in 1946. He went on eight successive tours by the British Lions, each followed up by a book. In all, he wrote 28 books on rugby between 1954 and 1980 and became a founding member of both the Rugby Union Writers' Club and the Welsh Rugby Writers' Association.

He was a well-known and trusted friend of many Welsh sportsmen and administrators, to an extent that on the Thursday lunchtime before a Saturday international, Bryn would often receive an invitation from one of his well-connected friends in the Club where, later in the card room, he would be quietly advised of the likely starting line-up for a forthcoming game. It was therefore with more than a touch of black humour about how well-connected JBG was, and the influence of his writing, that his editor once told him to "make a few mistakes" when predicting the Welsh XV so that the readers of the *Western Mail* would not think it was actually he who picked the side!

He was appointed MBE for his services to journalism in 1984 and, after retirement, he joined the Cardiff and County Club where he had been a guest of, among others, Sir Robert Webber, Wilf Wooller and Sir Tasker Watkins on a host of previous occasions. Despite in later years having a leg amputated and being confined to a wheelchair, he found plenty of time to mingle and reminisce with his many friends in the Clubhouse.

Had both Bryn and Henry Lewis been alive in July and August 1997, they would each have revelled in the lively chat in the bar about the forthcoming vote for devolution, as well as the arguments about a National Assembly in the Welsh capital. Chat about these topics did, however, cease for a while during early September following the tragic death in Paris of Her Royal Highness, Diana, Princess of Wales. Many in the Club had known and acted in an official capacity alongside the wife of Prince Charles and, as a mark of respect, the Club was closed on Saturday, September 6th, the day of her funeral in London.

The plans for the creation of a National Assembly building in Cardiff had received the green light following the passing of the Government of Wales Act 1998, which gained Royal Assent on July 31st. After inspecting a number of possible locations and suitable buildings, a site at Capital Waterside in Cardiff Bay was acquired and an international design competition was held. It was won by the eminent architect Richard Rogers whose plans for the Senedd embraced renewable technologies and energy efficiency. Building work began in 2001 and despite being well over budget, the Senedd was handed over to the National Assembly a few weeks before its formal opening by The Queen on St. David's Day, 2006.

The Senedd has subsequently become a symbol of modern Wales and, by linking activities with Tŷ Hywel and the Pierhead Building, at which the Bute Docks Company was based, the Assembly buildings cleverly embrace the old and the new – a theme which the members of the Cardiff and County Club took on board as they moved into the 21st century.

NOTICE
MEMBERS ARE REQUESTED NOT TO SMOKE IN THE DINING ROOM BEFORE 1.45 P.M.

'No Smoking' but did members ever fume?

26

The Millennium Stadium

"There's nowhere else quite like the Club."

When, during the 1880s, the members of the Cardiff and County Club first visited the plot of land adjacent to the Racquets and Fives Club in Westgate Street, they could not possibly have imagined the changes that would take place around the Arms Park a century later. In fact, I doubt if even the most forward-thinking, or drunk, member in the Clubhouse at its opening dinner in 1892 could have dreamt that 120 years later the area next to the Club would be the focus of worldwide media attention. But, with the arrival of the 1999 Rugby World Cup, followed by a series of FA Cup Finals, and the football matches in the 2012 London Olympics, plus the 2017 UEFA Champions League Final, what had started life as an area of river meadow straddling the Taff was now familiar to a worldwide television audience of millions.

The redevelopment of the stadium, further changes around the perimeter of the Clubhouse, and the emotive issue over lady members were the three major themes in the life of the Club during the closing years of the 20th century. As far as the first two topics were concerned, the redevelopment of the National Stadium first hit the headlines in 1995 as the Welsh Rugby Union won the right to host the 1999 Rugby World Cup. With the National Stadium starting to show its age, and with a capacity of just 53,000, the Union took the bold decision of replacing the Stadium with a modern and multi-purpose state-of-the-art facility. The eventual scheme cost £126 million and was part-funded by the National Lottery, as well as by ticket debentures and loans which left the Union in substantial debt.

It also led to a couple of years of careful politicking by the senior management at the Club as they shrewdly looked to preserve one of their most prized assets, namely the car parking facilities on both the access road adjacent to the Clubhouse and in the rear gardens, both of which gave the Club the invaluable advantage of being able to offer on-site parking

Lawrie Williams.

to members, diners and guests. The potential loss of these facilities by the redevelopment of the stadium, as well as during the construction phase, would have had a massive impact on the income streams on which the Club was so reliant. The news of the Union's plans came a few months after the Club's committee had discussed a financial shortfall of around £10,000 for 1994-95 and how, through more varied menus, more members could be encouraged to eat in the Club.

Following the publication of draft plans for the redevelopment of the stadium, the Club appointed a standing committee, chaired by Richard Morgan, and subsequently renamed the Advisory Group, which contained three members with the appropriate professional and business expertise who could represent the Club in any negotiations with the Union. Lawrie Williams, the Club's Honorary Solicitor, also joined the group with the Oxford-educated solicitor, and the fourth generation of his family to play a leading role with the Club, acting as a valued adviser.

Lawrie had first visited the Club on international weekends to sit alongside his father 'Tip' on the temporary seats installed on the roof of the Club. He duly sat enraptured on many weekends when Wales played in the Home Nations Championship, besides watching various events in the Empire Games. After reading law at Lincoln College, he joined the family's practice in Charles Street and later Windsor Place, and became a full member of the Club. Lawrie would often join his father – as well as acting as chauffeur afterwards – at prestigious events, such as the Centenary Dinner, although protocol meant that as a junior member he had to sit at one of the tables upstairs rather than being by his father's side as he addressed the gathering. Lawrie also regularly took lunch in the Club during the week, leaving his office shortly after 1pm before returning just before 3pm. He said: "Spending lunch in the Club and amongst the company of my friends, helped to clear my head and my work did not get bogged down as a result of spending hour upon hour poring over documents."

Given his family's links with the Club, and his many years of enjoying all the benefits the Clubhouse could offer, there could have been no finer person than the well-respected Cardiff solicitor to join the group. In the next couple of years, Lawrie's wise counsel prevented any major difficulties from arising and hampering either the Union in its lofty ambitions ahead of the showpiece rugby event, or depriving the Club of its prime asset. He also helped to build on the harmonious relationship which his good friend and fellow Club member Sir Tasker Watkins had helped to nurture through his presidency of the Union.

Indeed, it was Sir Tasker who informed the 1995 AGM that the plans which the Union were poised to submit to South Glamorgan County Council for planning permission would not adversely affect the Club. The following March, after the submission of the grandiose plans, Sir Tasker also outlined to the committee and the Advisory Group how the nearest new stand would be 30 metres further away than the one currently standing, and that it was likely to improve both light and air circulation at the rear of the Clubhouse.

But there were other matters to consider, especially the potential loss of trade when the demolition and construction work was scheduled to take place, the reimbursement of any professional fees incurred by the Club, and several matters relating to car parking. The latter related to maintaining access to the Club's car park during the building phase and ensuring that there was the continuation of what had previously been agreed with the Union should the estimate of 30 to 40 events a year at the new Stadium come to fruition. The first outcome was the creation of a new sub-committee to negotiate with both Millennium Stadium Limited and the WRU. Their first request to Millennium Stadium Limited saw a refusal to consider covering any professional fees incurred by the Club. Fortunately, the Union proved to be more helpful and an agreement of up to £7,500 was subsequently agreed in August 1997.

The relationship between Millennium Stadium Limited and the Club improved during the early months of 1997. This was at a time when a few amendments were required to the plans which had been submitted for planning approval, especially concerning the re-orientation of the new stadium so that an entrance in Westgate Street would be the main point of vehicle access. The Club held a trump card in that they had a right of way on the access road into the existing stadium which was now going to be the main entrance, so the new company offered unrestricted vehicle access, as well as 26 parking spaces on a 150-year lease, at a 'peppercorn' rent. When added to the same number of spaces on the Club's freehold premises, this meant that the number of parking spaces available to the Club could drop from 80 to 52. An olive branch was also offered through the possibility of a further 20 spaces at an unspecified time in the near future.

The potential loss of more than two dozen parking spaces raised the hackles of several members and, after further negotiations with the stadium company and the Union, the offer of 26 spaces was raised by eight. This led Jeff Sainsbury, the Club's Chairman, to call a special meeting at which the members would decide whether to accept a total of 60 spaces including those on the Club's land, and the prospect of 20 extra spaces at some time in the future, or to reject the proposals and run the risk of the Union terminating the existing arrangements when they ceased in 2001, which would leave the Club with a mere 25 spaces. Sainsbury said that "it is in the interests of the Club to be seen as supportive of the redevelopment of the National Stadium" and, after weighing up the pros and cons, the offer of 60, plus a possible 20 in the future, was accepted.

Nevertheless, discussions continued with the Union over the Club's rights of way and, with the stadium needing to be operational by the late summer of 1998, concerns were expressed in some quarters over the delay in reaching a suitable agreement. In September 1997 Lawrie Williams duly advised the committee that "the Club would not want to be seen to be responsible for holding up the construction of the new stadium by not agreeing to forfeit its right of access over land at the Westgate Street entrance." An agreement was reached soon afterwards, together with acceptance of the Union's offer of 56 temporary parking spaces, all free of charge, in the NCP premises beneath the Westgate Street flats. The Club's Trustees duly signed the paperwork and a formal agreement took place on December 23rd, 1997.

Demolition work had begun in May 1997 and, by the early months of 1998, the East Terrace had disappeared as the new stadium, with its re-aligned pitch, started to take shape. The Club's committee were delighted to note in May 1998 that the work by the various contractors had all been done without any serious problems for the Club. The anticipated loss in trade which they had feared during this initial phase of work was also not as great as first feared, with a steady flow of diners and other guests going into the Club.

However, Brunswick Construction dropped something of a potential bombshell during the late summer of 1998 when they told the Club they would require a longer period of construction than first anticipated and could not guarantee handing back access to the Club's car park by August 30th, 1999. There was a possibility that it would not be returned until October 2000, leaving the Club without their on-site facility during the period when the Rugby World Cup was taking place and when the Club would be operating at full stretch. Fears of a massive drop in potential income were allayed early in 1999 when the Union's solicitors confirmed that the car park would be returned by the following August. Soon afterwards, Taylor Woodrow began work on the land formerly occupied by the Job Centre in Westgate Street, and adjacent to Jackson Hall, to create a new access area up to one of the new rows of turnstiles leading into the revamped Stadium.

By the time the shell of the new stadium was created, two further prominent members of both the Club and the legal world of south Wales had died. In January 1998 both Leslie Shepherd, a Club member since 1951 and former honorary secretary of the Cardiff and District Law Society, and Judge Bruce Fletcher Griffiths, a Club member since 1953, died. Griffiths had been one of the most distinguished members of the judiciary in Wales, sitting on the Wales and Chester circuit from 1972 until his retirement in 1986.

Educated at Whitchurch Grammar School and King's College, London, Griffiths served with the RAF until demobilisation in 1947. He was called to the Bar at Gray's Inn in 1952 and, from 1964 until 1970 he was chairman of the Local Appeals Tribunal of the Ministry of Social Security in Cardiff, besides acting from 1968 to 1972 as vice-chairman of the Mental Health Tribunal for Wales. Before taking silk in 1970, he was deputy chairman of Glamorgan Quarter Sessions and Commissioner of the Assize Roll Courts of Justice in London.

A regular communicant at St. Mary's in Whitchurch, Bruce Griffiths was a member of the governing body of the Church in Wales from 1978 until 1992, and president of its Provincial Court from 1979 to 1992, besides serving as Chancellor to the Diocese of Monmouth. The Barry-born judge also played a prominent part in the cultural affairs of Wales, especially the visual arts. He was particularly enthusiastic about modern Welsh art, and he filled his home in Whitchurch with canvases and busts by contemporary artists. He was appointed to the Welsh Arts Council in 1972 and to the chairmanship of its Art Committee three years later. He spoke with authority and eloquence on behalf of visual artists and drew on all of his skills as a barrister to argue for a greater allocation of the council's funds, besides never missing an opportunity of ensuring that Welsh art was promoted at home and abroad.

Between 1981 and 1992 he was chairman of the Welsh Sculpture Trust and did much to encourage the art form. He also helped to advise the Club about both the value and merit of the items on display within the clubhouse but, from the general public's point of view, it was as a leading member of the Contemporary Art Society for Wales that Bruce Griffiths left the most lasting impression. He served as Chairman from 1987 to 1992, and thereafter as vice-chairman, overseeing with great aplomb the commissioning and exhibiting of work by living painters.

Through his work with the Welsh Arts Council, he was also instrumental in bringing a number of European artists to Wales and, in recognition of this work was presented with a silver medal on behalf of the Czechoslovak Republic in 1986. Bruce Griffiths was also well-read in English poetry, and every Christmas he took great delight in making a small anthology of his favourite poems which he sent to a select number of his friends from the legal world, as well as within the Club.

February 1999 saw a visit to Cardiff and the Millennium Stadium by Leo Williams, the chairman of the Rugby World Cup Committee, together with other key movers and shakers who were planning rugby's first world cup in the game's professional era. The visit on February 16th, saw several dignitaries hosted for lunch in the Clubhouse and, at the request of the Welsh Rugby Union – and Sir Tasker Watkins in particular – the World Cup flag was flown from the Clubhouse's flagpole alongside the Club's own flag to mark the special occasion.

On June 26th, 1999, after a phase of night-time construction work under floodlights, the first match at the redeveloped venue took place, as Wales recorded their first ever victory over South Africa. The stadium wasn't yet finished but during the late summer further building work saw its completion, in time for the launch of Rugby World Cup.

The prestigious international competition and other events at the stadium have brought a much-needed increase in trade and footfall across the Club's threshold. At a time when businesses in Cardiff and across south Wales were starting to feel the pinch of the economic depression, the influence of 'Cool Cymru' was permeating the hallowed walls of the Clubhouse. Besides being a highly visible statement about the importance of sport in the life of the capital city and across Wales in general, the creation of the Millennium Stadium, and a modern-day barometer about the state of the nation, has helped to bring a new lease of life for its more genteel neighbour on Westgate Street, thereby allowing the Club to move forward confidently into the 21st century and successfully ride a number of difficult years.

The Millennium Stadium: the visible barometer of the Welsh economy

The Millennium Stadium proudly hosted the opening ceremony of the 1999 Rugby World Cup and in the opening game Wales defeated Argentina. Just over a month later, on November 6th, it was Australia who defeated France in the final, and in front of a full house of 72,500. At the time, this was the largest crowd at a rugby match in the capital city. The new Stadium, incorporating Britain's first retractable roof in a sporting venue, was at that time the largest sporting venue in the UK.

Despite being overtaken in capacity by Manchester United's Old Trafford in 2006, the Millennium Stadium soon became a popular venue for other sports, including football. The first Welsh soccer international at the stadium took place in March 2000 against Finland,

Building work also continued at night.

The magnificent national rugby stadium, on the banks of the River Taff, widely regarded as the most atmospheric in world rugby.

and the redevelopment of Wembley Stadium enabled the jewel in the crown of the Welsh Rugby Union to play host to several major football finals, including six FA Cup Finals from the spring of 2001 until 2006.

Among the other major sporting events to have visited the Millennium Stadium are the Rugby League Challenge Cup Final, Grand Prix speedway, World Rally Championships, World Championship boxing and indoor cricket and, together with a variety of musical concerts, they have brought large numbers of new visitors into south Wales and the Welsh capital. The number of repeat visitors, and the fact that so many organisations and sporting bodies, including those involved in the organisation of the 2012 Olympics, have requested to use the stadium – renamed the Principality Stadium after a sponsorship agreement with the Welsh building society – bears testament to the fact that Cardiff is firmly established as a popular and well-organised venue for international sport.

27

Days of the Dinosaur

"Visiting the Club is a totally unique experience."

For many years, ladies had been welcome in the Cardiff and County Club only when either signed in for lunch as guests on Saturdays, at the various cocktail parties at night with their spouses, when attending other privately organised functions, or downstairs in the Jubilee Suite on Monday and Wednesday afternoons. One of the instigators of this relaxation of the strict men-only regulation was Mary Parry-Evans, the wife of Sir Lincoln Hallinan, who had taken great exception to not being allowed to accompany her husband to a Mess Dinner in the Club. It was a measure of Sir Lincoln's standing and diplomacy that within a year the rules were suitably amended.

From the 1970s onwards, society in Cardiff, as well as in Wales as a whole, saw sweeping changes to the role and identity of women, with female liberation and sexual equality becoming part and parcel of many other areas of life. On many occasions during the past 50 years, the topic of lady members had been raised in committee, but an increase in female participation or as members of the Club was seen by many as a taboo subject, especially by those for whom being part of a gentlemen's club had been an attraction for joining in the first place. Others took the middle ground and expressed no strong feelings either way, while many were quite happy with the *status quo* as long as they could have a good time, chatting, eating and drinking with their pals.

A few argued that the Club had existed very well since 1866 without lady members so why change now? Others, who were very *au fait* with the legal position, argued that, as a private members' club, they were not doing anything illegal and suggested instead that the ladies were perfectly entitled to form their own club or group, and to meet regularly elsewhere.

But for a group of members from the late 1980s onwards, the exclusion of women from the membership was viewed as the Club's Achilles heel and something which needed to be amended, especially as there had been a female Prime Minister in the UK since 1979, whilst many women held senior positions in industry and commerce. To the minds of this group, the male-only policy of the Club brought adverse and unwanted publicity – for a few in this group, the drive towards including women in the membership became something of a *cause célèbre*.

A series of events during the late 1990s saw the arguments for and against women's membership being played out in the media. Though those in the pro-women camp believed that such a high-profile campaign might work in their favour, what transpired was a greater polarisation of views; and the more the newspapers called for change, the more the heels were dug in, with senior members of the Club refusing to change their position. As a result, the Club retained its traditional male-only policy for another decade at least.

What prompted the arguments for an end to the male-only policy was a series of news stories in local newspapers when a group of women members of the National Assembly Advisory Group (which drew up standing orders for the new body) decided not to enter the Club for a celebratory dinner on July 24th, 1998, hosted by John Elfed Jones, the group's chairman, to mark the end of their work. As it was a private function, they were entitled to attend, but some of the six women on the 14-strong group declined the invitation as a matter of principle with Eluned Morgan, the Mid and West Wales Member of the European Parliament claiming, in an interview with the *Western Mail*, that: "It was the first time since I have become an MEP that I have felt excluded because of my gender. There was no way I could go."

Ms Morgan also levelled criticism at Club member David Rowe-Beddoe, the chairman of the Welsh Development Agency, claiming that as he was paid from the public purse he should live up to the equal opportunities standard set by the WDA. "As a member of the Cardiff and County Club he, along with the other members has the power to change the discriminatory membership policy," Ms. Morgan said. "I urge him and the other members to use their influence to create a member's policy which acknowledges the rights of women."

Another to refuse the invitation was Helen Mary Jones, at the time the deputy director of the Equal Opportunities Commission for Wales and the equal opportunities director on Plaid Cymru's national executive. "It was unfortunate and a slightly curious choice," she said in an interview with the *Western Mail* when asked about the choice of venue for the celebratory dinner. "It was particularly depressing as we had spent so much time in the group talking about inclusiveness and exclusiveness and why it was necessary to have more women not just in the Assembly but in public life in general."

However, the newspaper's coverage of the topic also came in for criticism by some of its readers, with one gentleman from Pontymister complaining how an article had 'described the Cardiff and County Club as one of the last bastions of a men-only members policy, implying that such a policy is almost morally wrong since it allows men the company of other men, which is presumably why the members go. Clubs are designed to cater for a particular section of society – youth clubs for example, would you accuse these of ageism? If these so-called democrats were consistent in their wolf-crying of sexism, they would be proposing the abolition of the WI or *Merched y Wawr*. Of course, that would be completely unreasonable, but to slag off a men's club is easy prey, because any retaliation can be dismissed as chauvinism.'

But the *Western Mail*'s sister paper, the *South Wales Echo*, continued to beat the drum for the pro-women lobby, and on August 7th, Dan O'Neill wrote in his column – entitled 'The Kairdiff Kid' – as follows: 'So we've got the old Arms Park on one side of the fence and clubland's equivalent of Jurassic Park on the other. Jurassic Park, of course, is where you find strange lumbering beasts who have no place in the modern world, oddly-shaped creations that for most of us are part of a dim and oh so distant past. Just like the Cardiff and County Club. This august institution clings to its men-only membership policy and in other ways, as well, has yet to be dragged into the 20th century, even though the 21st century is only months away.'

For those who were undecided on the issue, it was a source of embarrassment to see the Club being ridiculed in this way. During the next few months Sir Geoffrey Inkin, the chairman of the Cardiff Bay Development Corporation, who was also the Club's Chairman, tried to smooth a few ruffled feathers by raising the issue once again at committee.

Indeed, Jeff Sainsbury, one of his predecessors as Chairman, had drafted a paper as recently as January 1997 to discuss the question of female membership following comments made to him by some members, but at the Annual General Meeting in April, it was agreed that no changes would be made to Club policy. In 1998 another Club member, a high profile figure in the media, also wrote to the Chairman expressing concern about the way the Club was being publicly held up to ridicule for not admitting women into membership, as well as drawing attention to how in families where both parties had professional careers, the existing rule on membership was viewed as being unacceptable to both parties.

Depending on your stance with this particular argument, it was either fortuitous or unfortunate that in September 1998 the Marylebone Cricket Club whose home is at the historic ground at Lord's and is regarded by many

Jeff Sainsbury, with his wife Jan, on the Clubhouse balcony.

as a bastion of male chauvinism, voted to admit lady members. Their president, Colin Ingleby-McKenzie, had written in a missive to members that 'it would be very difficult to find outside finance for MCC's external activities while the club excludes women from membership.'

It should not though be forgotten, that there were – and are – very important differences between the Cardiff and County Club and the MCC, in that the latter had a wide range of activities all over the world for which it needed funding. Rather than just being a dining club and a refuge for business people in a solitary location, the MCC also adopts a governance role worldwide, as well as looking to nurture and promote the playing of cricket. For these reasons alone, unlike the Club in Westgate Street, having both a secure and substantial funding stream was of paramount importance. Moreover, the MCC had also for several years already awarded honorary life membership to women.

Nevertheless, the news that the MCC were admitting lady members came as music to the ears of those lobbying for change, with Eluned Morgan saying that "if the MCC can do it, then so can the Cardiff and County." Their campaign also gained the support of two prominent women in public life in Wales – Glenys Kinnock, the former MEP and wife of Neil Kinnock, plus Menna Richards, who then was the managing director of HTV Wales. Mrs. Kinnock wrote to Sir Geoffrey Inkin asking for further information about the Club's policy on women, and Ms. Richards claimed that she had the support of two existing Club members who were ready to propose and second her membership. Sir Geoffrey also took part in an interview on BBC Radio Wales with Glenys Kinnock as the topic got another airing on national radio.

David Crosby was one of the members who believed that the time had come for a change in membership policy and during early 1999 he made it clear that he would be proposing at the AGM in June that Rule Six should be amended by substituting 'person' for 'gentlemen.' The thought of further heated debate at the annual gathering did not appeal to many, especially Sir Geoffrey Inkin and, after further discussions, Crosby agreed to withdraw his proposal on the understanding that some form of consultation would be held, hopefully to allow informed discussions to take place rather than a potentially divisive debate. With this in mind, the committee agreed that they would recommend to the AGM that an independent postal vote would be held, prior to a special meeting in September, to provide 'an authoritative indication of the views of the widest possible membership.'

Their suggestion backfired as it failed to gauge the mood of the majority of members at the AGM, who were quite happy with the *status quo*. It also brought into sharp focus the role of the committee and the governance of the Club's premises in Westgate Street. Indeed, after Sir Geoffrey had

Sir Tasker Watkins at the unveiling of his portrait, with David Griffiths.

presented his Chairman's Report and raised the issue of an independent postal vote, Sir Tasker Watkins – in his role as a Trustee – queried the actions of the committee and whether they were being bullied by the media and a minority of members into making changes.

"There is no power for the committee to have a ballot as suggested," he said, before adding: "as it currently stands, this is a most valuable Club, which is highly regarded all over the world. This is a gentlemen's Club and nothing else. The committee may not do what they seek to do without the members' support. It should not be moved by the media to do anything. There is nothing wrong with the rules as they stand. The Club is properly constituted. Anybody can propose that this or that rule be amended, but the rules do not permit a trial run. Power reposes with the membership not the press. Freedom to assemble is a cornerstone of our law."

Sir Tasker ended his impassioned plea by advising members not to support the proposition of a postal ballot. Sir Geoffrey duly responded that there was no intention of undermining the rules of the Club, but instead to provide information for a future meeting. Roger Thomas, one of the other members who had written to the Chairman suggesting that ladies be admitted, leapt to the committee's support by saying "society has moved on. It is proper to consider whether there should be a change."

Lawrie Williams, as the Club's Honorary Solicitor, duly reminded everyone that it was not necessary to vote on the Chairman's Report, but merely to receive it and note the comments it contained. Glynne Clay, another Trustee, then proposed that the report be received but, aware that a postal ballot would mean that the matter lingered on, he asked members for a show of hands to indicate if they wanted to change the rules to admit lady members. The result was that only a dozen of those present were in favour of a change compared with an estimated 90 against any changes.

It had been one of the most contentious annual meetings ever held at the Club, and the following week Sir Geoffrey Inkin resigned, both as Chairman and as a member of the committee. "In view of the implicit lack of confidence of the membership in the committee, it is with the greatest regret that I have no option but to resign," he said at the meeting on July 5th, 1999. "The acrimonious debate at last week's AGM on an issue which had not been submitted formally, illustrated only too clearly the latent potential for divisiveness and disagreeability. The committee had been particularly sensitive to the danger and had devised an alternative to provide an indication of the wider membership's view. Ironically, this consultative process would not have been constitutionally different to the impromptu show of hands at the AGM."

David Watson James was duly elected in his place, and at the committee's next meeting in August it was agreed that the Chairman's Report had proved to be a flawed vehicle for a recommendation which required such an important decision and on an emotive issue which struck right at the heart of the spirit and ethos of the Club and its members. Not surprisingly, the committee agreed that no further action should be taken, but the matter still rumbled on outside the confines of the Club with a school of thought being that by providing opportunities for informal networking among men of influence and power, the Club was operating against the wider interests of broadening equality of opportunity for women.

David Watson James.

In October 1999, the Club declined an invitation to take part in a programme on BBC Wales about clubs in Wales, while Val Feld, the Labour AM for Swansea East, was active in the media criticising the Club, and arguing that no public money should be spent on the Club or its premises. Though the BBC Wales programme was not too damaging about the Club, the lobby against the Club and what it stood for took a different turn with adverse media comment in January 2000 about the Clubhouse having 13 pictures on loan since 1949 from the National Museums and Galleries of Wales.

Following the publication of these articles about the on-loan items, the Honorary Secretary and General Manager each received telephone calls from members of the public who wished to view the pictures. The requests were declined, as there was no mention in the 50-year-old correspondence that the pictures loaned to the private members' club should be made available for public viewing. The Welsh Office then weighed in by adding that 'it is not appropriate that the pictures should be lent to a body that does not practice equal opportunities.'

Given this adverse publicity, it came as no surprise to the committee and many members that in early March the council of the National Museums and Galleries of Wales decided that the 13 pictures on loan to the Club should be returned. David Watson James had an amicable meeting with Anna Southwell, the manager of the Museum, before the paintings were duly removed from the Club's premises on June 20th, and all without the blaze of publicity some had expected.

For a short while there were some bare spaces along the corridors and on the walls of the Clubhouse following the removal of the pictures, but there was a positive spin-off as far as the Club was concerned. As David Watson

James later explained: "Many of the pictures had rarely been looked at by our members, so there was not a huge uproar amongst the membership when they were returned to the Museum. In fact, their departure prompted us to look carefully at the remaining collection and with this in mind, I created an Art Committee and an Art Fund, to help rationalise the collection and improve the décor."

It was not long before a number of suitable and grandiose replacements were secured, with a host of favourable comments being passed on to the Chairman and the Art Committee about the new pictures. What could have turned into an almighty upset for the Club had been adroitly handled by the Chairman and committee. The Club and its members had stayed true to their principles and remained steadfast in their beliefs. To some, it only hardened the resolve to remain a male-only Club and, as one senior official said, when asked about the position by a journalist, "we have not done anything illegal. This is a private Club and it is nobody else's business!"

The Art Committee

After its creation in 2001, the new group was chaired initially by David Williams, with Professor Sir William Asscher as Treasurer, David Crosby as Secretary, and Dr. Peter Wakelin and Keith Mainstone also taking part. Their remit was as follows:

- to prepare a strategy for the acquisition of paintings and other suitable works of art to be hung on the walls of the Club's premises;
- to acquire suitable items that would enhance the ambience of the Club;
- to dispose of such works of art which were deemed redundant;
- to maintain a list of works of art, together with their value; and
- to oversee the care and conservation of items.

The audit showed that the collection comprised many watercolours, drawings and etchings of scenes in and around Cardiff, various works on aspects of the history of south Wales, portraits of eminent members and officers, including a loaned oil painting of Sir Cennydd Traherne, a variety of sporting and rustic prints showing a number of country sports, including some John Leech hunting prints. However, there was agreement that the outstanding work in the Club's collection was Denys Short's 'Porth, looking towards Pontypridd' which had been donated by Kenneth Davies CBE, who had bought the painting at the 1960 National Eisteddfod.

28

Lift Off! Into the Twenty-First Century

"It may have been formed 150 years ago, but the Club is still an important, and relaxing, place in the hurly-burly of life in cosmopolitan Cardiff."

It was with a huge sigh of relief within the rooms of the Cardiff and County Club that the construction of the Millennium Stadium did not damage the fabric and heart of the establishment in Westgate Street. Indeed, the Club's newly-completed neighbour went from strength to strength during the first decade of the 21st century, successfully staging a series of major sporting events such as the FA Cup Final, which helped to regenerate many aspects of the region's economy, as well as reinforcing the name of the Welsh capital as a city of global sporting importance.

What was good for Cardiff and the Millennium Stadium was – in theory – very beneficial for the Cardiff and County Club, with the potential for increased takings and new opportunities for hospitality as well as membership. After the heated debate at the 1999 AGM, which was followed by newspaper and radio articles about the exclusion of women as members, the last thing that the committee wanted was for the topic to erupt again, and this time to the detriment of the Club.

Although some outside the walls of the Cardiff and County Club still believed that, by excluding women members, the private members' club was becoming increasingly out-of-date and projecting values irrelevant to a modern capital city, inside the premises of the men-only establishment, there were several developments during the early 2000s to show that the Club was moving into the modern era. Within a few months of the Clubhouse being bestowed with Listed Building status, arrangements were made for the installation of satellite television, with Sky TV being available for members from mid-November 2001. By February 2002, members could also communicate with the Club's officials by email and, from October that year, the Club had a website plus online credit card facilities.

As well as these changes to facilities, there were also some important changes in personnel. Ill health forced Derek Tantum into taking early retirement from September 30th, 2001, with Major Nicholas Cooke taking over as General Manager. February 2001 sadly also saw the death of Sir Melvyn Rosser, a well-known Club member who had formerly been chairman of HTV and president of UCW Aberystwyth. Sir Melvyn had been a Club member since 1967, and during his long and distinguished career in business and public life, he had been non-executive director of the British Steel Corporation and the National Coal Board, besides being chairman of the Manpower Services Committee for Wales between 1980 and 1988.

The longest-living member?

March 2001 saw the Cardiff and County Club lose one of its longest-living members with the death of Albert John Alexander, just a handful of weeks after celebrating his 100th birthday with a special dinner in the Clubhouse. Albert had spent a lifetime working at Cardiff Docks as a chartered shipbroker with T.D. John (Shipping) and during the Second World War he had managed vessels for the Belgian government after their country was occupied by the Germans. His painstaking efforts were rewarded in 1948 by the presentation of the order of *Chevalier de l'Ordre de Leopold II*.

In subsequent years, Albert had played a pivotal role in the twinning of Cardiff with Nantes and the creation in mid-December 1973 of the Cardiff-Nantes Fellowship, a voluntary organisation under the patronage of the Lord Mayor of Cardiff with the objective of advancing mutual co-operation of knowledge and understanding between the people of the two cities. Annual exchanges and reciprocal visits duly took place each May and June, with Albert taking great delight in hosting French dignitaries as his dining guests in the Clubhouse.

Within a year or so of the completion of the stadium, it became apparent that it may have had an adverse effect on the Club, with the number of functions and members using the Clubhouse taking something of a dip during the early 2000s, especially when Westgate Street was closed because of various events within the stadium, particularly on Sundays and on other evenings. The drop in the number of diners was also a reflection of a drop in membership and, as the monthly takings started to show a noticeable downturn during 2001, the Club's Management Committee spent several meetings discussing the potential loss of £20,000 for the year.

The outcome was the establishment of a Development Sub-Committee which began work on creating a blueprint for the way forward. Among the issues discussed were the effective marketing of the Club's facilities, and the re-allocation of areas which, at the time, were not generating revenue, including the billiard room and other areas on the upper floors. The introduction of lighter meals on the menu, as well as Sunday lunches were other items to be raised, together with the introduction of lady members, which many still believed was the Achilles heel of the Club as it moved into the 21st century. 'Serious consideration has to be given to admitting lady guests and the provision of appropriate facilities,' was one of the resolutions which the sub-committee agreed and, in February 2002, it was recommended that the billiard room should be used as a mixed gender function room for use at weekends and evening functions.

After the events of the previous couple of years, and the adverse publicity from some sections of the media, the last thing the Club wanted to do was to re-open old sores, so discreet soundings were taken with senior members in the ensuing year. This resulted in a memorandum from the Chairman, Bob Edwards, who on May 29th, 2003 stated that: 'there is clearly a strong body of opinion within the Club which would resent ladies attending the ground floor of the Club during weekdays and I believe that view should be respected. Consequently to particularly satisfy the need of members to host lady guests and to avoid upsetting the members who wish to have exclusive male use of the ground floor, I propose that we allow lady guests accompanied by their host into the Jubilee Suite for dining purposes on Thursday and Friday subject to giving the Manager at least twenty-four hours' notice.'

Born and bred in Penarth, Bob had read law at Aberystwyth before practising as a solicitor, specialising in family law. An avid golfer and former captain of the Glamorganshire Golf Club, he twice served as Chairman, from 2002 to 2004 and again from 2011 to 2013.

His well-chosen words during the discussion on lady members headed off any prospect of a further storm within the corridors of the Club, but a more pressing issue was the question of disabled access to upper floors and the installation of a lift. It was prompted by the Disability Discrimination Act of 1995 which made it a statutory requirement for lifts to be installed, although private member clubs were exempt if upper floors were not used for commercial purposes. Many believed that the installation of a lift would result in

Bob Edwards.

the vast room space on the upper floors being properly exploited, leading to the structural engineers drafting their plans and displaying their ideas for members to inspect.

Some quite radical suggestions were also put forward, including a partition in the lobby and other major internal works which several members described as 'architectural savagery'. They argued that the building being old-fashioned was a virtue, not a vice. In the words of one distinguished member, "the building has great character and this is amply reflected in the warm-heartedness and generosity of its staff over many years, with a lineage of kindly waitresses and bar staff from Mary and Betty to Edna, Sally and Dawn."

The redevelopment plans comprised offices on the top floor, dining and catering facilities on the first floor in the billiard room, the conversion of the committee room into a reading room and light dining area, plus the installation of a lift to all floors – all at a cost of £250,000. With the input of General Manager Nicholas Cooke, it was an attempt to transform the Club – still seen by many as principally a luncheon club – into a far more commercial organisation, especially with plans for an up-market, first floor restaurant where people would also dine in the evenings. In presenting these ideas, the committee added:

> 'the Club should move forward by utilising existing members' resources and enhancing its facilities which, in turn, will help promote greater interest in, and the use of, the Club …. The existing operations use approximately one half of the Club's property and, unless the unused and underused portions of the building are brought into productive use, the cost of their maintenance will eventually become unsupportable and will undermine the financial stability of the Club's operations. The Club currently operates on a sound financial basis and it is essential that we move positively and proactively into the future, providing those facilities which will be required by future generations of members.'

These were very laudable reasons, and out of the 190 replies which the Chairman had received to a mailshot outlining the ideas, 185 had supported the notion of redevelopment. But despite this positivity, the main bone of contention was the likely cost, and also the disturbance to the internal fabric of the building. Many felt that both were unnecessary, and some wondered if the greater use of the first floor was another means by which the pro-ladies lobby could win the day. To address the concerns over the financial issues, the committee reported that a bank and a building society had each been sounded out about a loan spanning 10 to 15 years.

The Club's bar doing brisk trade.

But the absence of either a development or marketing plan made many people quite nervous, especially given the size of the sums concerned and the length of indebtedness to which the Club would commit itself. The lack of any financial information, outlining how the capital expenditure and revenue generation of the Club might change over the coming years concerned many members, especially at a time when the local economy had taken a buffeting. There had been a rise in the number of vacant premises in the adjoining streets and, as the 'For Sale' boards became more visible in the city centre, many members of the Club realised that they could ill afford to make hasty decisions about its future.

On December 17th, 2003 the thoughts of many members turned away from the development plan and towards their President, as news came through to the Clubhouse that John Cory had died at the age of 75. His death in office cast a shadow over the festive season at the Clubhouse. Shortly before Christmas 2003, there was some good news for the Club and its members as it was confirmed that the Club would be receiving a VAT refund of some £150,000. When the committee met in early January, they agreed that this windfall would cover a large chunk of the redevelopment scheme. To meet the rest of the costs, discussions began about securing a loan of £100,000 as well as the release of about £50,000 from the Club's cash reserves of £95,000.

To assist with the planning of the proposed changes, tenders were also issued to four contractors, but further concerns were raised by members about the plans for the re-fit as they went around the building with the architects on January 7th. In particular, the plans for a partition and screen in the entrance lobby aroused a number of negative comments, as did the proposed location of the lift. Others queried the need for expenditure on additional dining facilities at a time when the existing rooms were underused. Staffing costs had already risen by 41% in the period from 1997 to 2003, and with more space to be filled, a further increase in staffing costs was predicted.

Some pointed to the fact that the Club's income during the early 2000s had been bloated by extra functions at the Millennium Stadium following the redevelopment of Wembley Stadium, with marquee events being shifted temporarily from London to Cardiff. Had it not been for this supplementary income, the drop in the Club's revenue would have been much greater. Others argued that the annual costs of maintaining the Grade Two Listed building stood at about £25,000 and that, as the years progressed, this sum was likely to increase. These financial concerns, plus the lack of a robust marketing plan, led the committee on February 26th, 2004, to scale down the scheme and to remove the plans for dining facilities on the first floor, besides putting a cap on the entire project to a maximum of £200,000, funded by the VAT windfall and a chunk of money from the Club's reserves.

There were also discussions in both the committee room and around the Club about whether another solution, albeit a very radical one, would be to vacate the Clubhouse and move elsewhere. One suggestion was moving to the adjacent County Court building which was now disused,

Three generations of the Jones family at The Old Cliftonian Society's Dinner, held in the Westgate Room at the Cardiff and County Club in October 1990. Back row (left to right): Nicholas Jones, Simon Jones and Andrew Jones. Front row: David Mansel Jones, Frederick Jones (father of David and Philip) and Philip F. Jones (now resident in Canada). The Old Cliftonians have held their annual dinner in the Club since the late 1970s.

while others suggested seeking more modern and purpose-built premises in the city centre or Cardiff Bay. But there were few sites which could offer ample car parking, and while the market value of the Clubhouse stood at £975,000, this would not provide sufficient funds to buy other premises, meaning that any move to other premises, including to the County Court, would be far more expensive. Similarly, the Club was not in a position to make a reasonable bid for the block of land known as 'Westgate Plaza' adjacent to the car park when it came on the market during the spring.

A Special Meeting was duly convened in the Clubhouse on April 29th, 2004 at which 104 members were present, with matters being overseen by Vice-President, Glynne Clay. After a minute's silence in memory of John Cory, Chairman Bill Gill outlined the committee's proposals as well as stressing the need for change. A number of the arguments against the plans were given a further airing, while Keith Mainstone, a long-standing member and an architect of many years' standing, put forward alternative ideas for redevelopment. Paul Twamley, the well-respected group finance director of Hyder PLC, also raised concerns about the use of the Club's cash reserves for the redevelopment fund when there was no plan in place for the recruitment of new members.

Like the meeting a few years before about the admission of lady members, there was opposition to the committee's suggestion and, though it was not as contentious a meeting as before, the proposals were nevertheless defeated by 61 votes to 43. It also did not result in any changes in management, and instead saw the committee take on board Messrs Mainstone and Twamley's comments, with each preparing papers for the committee to digest. The latter in particular felt that the committee had neglected to look at the issue of car parking, and 'unless this issue is faced now, the future of the Club is bleak. We should charge all members a suitable car park fee of £200 to £300 per annum and redeemable in the dining room. Those who join the Club for its parking facilities will be priced out, or even better, be persuaded to have lunch in the Club!'

Keith Mainstone.

29

An Agenda for Change

"You can dine here alone, but you are always amongst friends."

'Since the Club's foundation, our daily lives have changed beyond recognition. There are now very few gentlemen of private means, provision for horses at the Club is now rarely necessary and the motor car presents ever increasing demands. Ladies are not only both seen and heard, but also vote and work. Very few younger women are content with a life 'at home' and ladies are welcome guests in the Club at certain times, every day. The Club has changed many times, in the past, as its needs have changed and the committee has decided that this is the time for further changes.'

The words of Club Chairman David Watson James in his introductory notes to the consultative document, *An Agenda for Change*, which was sent to all Club members in November 2004. The document prepared by David and his committee reflected the feedback to the redevelopment plans, as well as the ideas floated by Keith Mainstone and Paul Twamley, plus the work of the Membership Sub-Committee chaired by Alun Davies and involving Dr. Gareth Rees and David Crosby.

The document also showed to hardened critics of the Club that the members were not as stuffy and introverted a collection of old dinosaurs as some would like to think. There had been quite radical developments across the way at the Arms Park, with the creation of the country's first covered sporting stadium. In many other parts of Cardiff, not least the Bay itself, there had been some sweeping changes and, as the document outlined, now it was time for change at the Cardiff and County Club.

The consensus was that the original redevelopment proposal, especially for the first floor, was too much too soon, and that the main priorities were to encourage people to use the Club, especially in the evenings, to boost membership and to establish a firmer business footing.

The quintessential character (and characters) of the Club

While the committee and officers were completing changes in the Club, great care was paid to any suggestion that would preserve the quintessential character of the Club, neatly summed up by David Watson James as "a place where convivial conversation can take place in a congenial atmosphere away from work."

Indeed, one of the delights of walking into the Club for lunch is that you never know who you might meet, and possibly dine with, as the mix of people from all aspects of life in Cardiff is one of the unique features of Club life. There are the regular groups such as the 'Debs', all of whom were sons of long-standing members from the pre-1939 era, who themselves joined during the post-war era. More than 50 years later, they still dine regularly on Fridays, besides gleefully revelling in a game of spoof to see who should pay for the wine.

The lack of political affiliation or allegiance is another enduring feature of the Club. As David Mansel Jones recalled: "There was one occasion when I came into the Club and saw, in the dining room, a table containing Sir Maynard Jenour, who was a staunch Liberal, Peter Temple-Morris, the Conservative MP for Leominster, plus a group of people from Plaid Cymru."

Members gather at the bar during the Club's 140th anniversary dinner.

Quite rightly, there are some rules about the use of mobile phones in the dining room, and smoking being confined to outside on the balcony, but there are some idiosyncratic ones, such as passing drinks through a hatch from the bar, as it is deemed unseemly to walk into the dining room carrying a pint or wine glass. Similarly, wearing or carrying a coat into the bar is frowned upon but once there spending time over a drink with fellow members affords the opportunity to meet and listen to a number of interesting and erudite people. Indeed, Robbie Norris still remembers the many times he mingled in the bar with Judge Hywel ap Robert – "a man who was fluent in seven languages, including Welsh, and someone with a mellifluous voice, plus an easy-going and warm-hearted nature."

There are certain idiosyncrasies, such as the Club's Cork dry gin-infused rice pudding and a host of larger-than-life personalities who brighten-up many a dull and cloudy day (and night) in Cardiff. For example, on one late afternoon during the long, hot summer of 1976, Wilf Wooller discovered that the Club's Manager had decided to fill up one of the baths on the upper floor in case there was a period of water rationing. Who else but Wilf could have then headed across to Cardiff Market to purchase some goldfish which, to the shock of the Manager, spent the next few days swimming around in the bath.

But even Wilf's jolly efforts pale into insignificance compared with those of Freddie Mathias, who devised a light-hearted form of hoopla, whereby he and other members – including some distinguished gentlemen from the ecclesiastical and legal world – could stand at the top of the staircase and challenge each other to see who could successfully lob bowler hats on to the horns of the moose head mounted above the door leading into the dining room.

His son Tim has retained the family's tag as being the Club's jester, and on international days Tim can be seen gleefully parading around the Clubhouse

Tim Mathias wears his patriotic Welsh 'uniform'.

in a smart red blazer – obtained from the Avis Car Hire company whose employees wore a bright red-coloured uniform – on which he has attached a Welsh Regiment XV badge.

However, Tim's most famous jape came in the autumn of 2014 when the Club's members assembled in the Clubhouse to vote on admitting ladies as members. It had been a thorny issue for many years, but a potentially difficult situation had been adroitly defused as Tim, dressed entirely in women's clothes and immaculately made up, calmly walked in and joined the gathering. As a hush descended in the room, Ceri Preece, the Honorary Secretary, approached Tim and said "Good evening, Tim. You know what I am going to say. You are not complying with our dress code." Ceri then paused for a few seconds before adding, "You are not wearing a tie!"

The Club's 'First Lady'? How Tim Mathias dressed for the special meeting to approve the introduction of lady members.

An Agenda for Change was a synthesis of ideas and other sensible suggestions which, the officials hoped, would act as the blueprint for taking the Club into the 21st century. With some well-crafted and diplomatically phrased sentences, *An Agenda for Change* also addressed the issue of a declining membership. Indeed, this was the opening theme of the document, which began as follows – 'the oxygen of success for any Club is a thriving membership and they play a key role in promoting the Club via personal contact and persuasion. The declining number of full members, coupled with the bad publicity in recent years over the issue of membership for women, has put the Club on the back foot. This is an appropriate time to move forward by attracting new members.'

A Membership Sub-Committee had been created to discuss a number of initiatives to attract new members. Chaired by Bill Gill, the sub-committee decided that a number of incentives needed to be made, especially as many directors and business leaders were less likely to go out, as before, for a relaxed lunch, with many professionals opting instead to pay a subscription to a golf club or gymnasium complex. Having an introductory offer of membership at just 75% of full fees was seen as one way of attracting people, and having enjoyed the ambience and met new friends, the hope was that they would pay the full amount in subsequent years.

The need to attract younger members was another important issue, especially as just 2% of the 680 members were younger than 40. Among the ideas to address this matter were a combined father and son membership at 150% of the current rate, as well as having a zero joining fee for anyone under 30, and reducing their subscription rate from £296 to just £100. It was also suggested that a working group could be created who 'talent spotted' new members, whilst another idea was to hold a cocktail party where each member brought as his guest a prospective new member. Other suggestions included simplifying the joining procedure and it was also mooted that members could receive a 10% return of their subscription if they acted as the successful proposer of new members. Changes to the distances and prices for the various categories of country membership were also discussed.

The sub-committee also revisited the issue of lady members and felt that there was still mileage in the original redevelopment plan of converting the committee room into a mixed bar and brasserie serving light meals. They also recommended publicising the Ladies Cardiff and County Club which met in the Jubilee Suite, and welcoming widows of deceased members, by inviting them to join the other ladies in the basement room. Indeed, in reporting to the full committee they noted that 'the ethos of a single sex club is less appealing now than thirty or even ten years ago, with the single sex policy having caused an unknown number of senior individuals to decline membership or simply to stay away.'

The re-visiting of the issue of lady members was the main virtue of *An Agenda for Change* and, with both sides having taken something of a less aggressive stance, it allowed the people of influence within the Club to spend time carefully, and without vitriol to discuss the situation and the changing social landscape of south Wales following the passing of equality legislation. Indeed, the responses to the consultative document showed a small majority, 52% to 48%, in favour of admitting ladies as members.

In all, 447 of the 685 members responded to the survey with 124 strongly agreeing to ladies being admitted as members, 99 agreeing, 89 disagreeing and 117 strongly disagreeing, with 18 spoilt ballots. This consultation exercise showed that more people were now in favour of changing the membership, but there was not – as yet – a clear majority in favour of lady members.

This was in stark contrast to the 96% who voted in favour of refurbishment to the furniture and fabric on the ground floor, as well as the 88% who had voted in favour of installing a lift that would run from the basement to the top floor, in addition to the creation of a new entrance to the Club from the car park. But, while there was agreement over the principle of having a lift installed, there was much greater concern over the precise detail and location of the equipment.

Alternative proposals had been floated as Paddy Gallagher and Tim Mathias submitted an alternative plan for a lift which needed neither a pit nor a motor room, and comprised level access walkways to the existing side entrances, besides not disrupting any of the internal fabric of the old building. Their scheme allowed for an external lift shaft by incorporating two additional levels and adding a further 43 car parking spaces. Another, quite innovative idea was to use the top deck of parking for hospitality on international weekends, and run along similar lines to Twickenham's West Car Park.

Paddy Gallagher.

As 2005 unfolded, and further discussions took place with Oakley's over both their plans and estimates for the cost of installing the lift, more concerns were raised. When their tender, amounting to £238,000, was duly submitted to the committee, the management group agreed in April 2005 not to proceed any further and to reassess precisely what was required.

Further initiatives to increase the number of members were introduced in 2005, such as the establishment of a Literary Circle. At the suggestion of David Watson James, important amendments were made to the membership procedures, with the number of proposers and seconders being reduced from four to two, while seconders had only to have been a member for at least three rather than five years. Professor John Lazarus and Iain Breckenridge also oversaw the establishment of the annual *Burns Nicht* and in the next few years more themed evenings were held, together with monthly Sunday lunches.

The questionnaire responses had broadly supported the notion that ladies should be admitted as luncheon guests on all days of the week, although there was still an element of resistance towards allowing lady guests to eat in the main dining room. The creation of an alternative dining area in the committee room was viewed as a solution to this impasse, with an alternative menu with lighter lunches also being available in the committee room.

2005 also witnessed several changes to staff and management with Mair Forsdyke retiring at the end of March. Ill health also forced Commander John Payn to stand down as Hon. Secretary, and Lawrie Williams retired as Honorary Solicitor in mid-April, with Adrian Larby taking over as his replacement. Arthur Weston Evans retired as the Club's Archivist in October, with the honour of Life Membership being conferred on him in recognition of his long service and authorship of the Club's 125th anniversary year history. John Cosslett, a well-respected and long-serving journalist with the *Western Mail*, took over the duties of Honorary Archivist.

It seemed as if changes might soon be in the offing. With plenty of the talk in the committee room focused on the results of *An Agenda for Change*, and an action plan for new developments being discussed, along with the arrival of new blood with potentially fresh ideas in the administrative affairs of the Club, Michael Jones was appointed in Commander Payn's place and, by the end of the year, the new Honorary Secretary was embroiled in quite lengthy and complex discussions with the ladies' bridge group who, since the mid-1970s, had been meeting twice a week in the Jubilee Suite to play bridge. In particular, he felt that their annual rent of £3,500 was insufficient, with a figure three times this amount being discussed as an alternative.

John Payn.

There were many who felt that such a dramatic increase in the rental arrangements for the ladies' bridge group would become a sensitive issue and lead to unnecessary animosity or claims of misogyny. Indeed, as some had predicted, the Honorary Secretary of the ladies' group contacted the Club's committee and argued that the proposed increase 'for an unfurnished room used two afternoons a week is excessive, especially when the group spends on average £350 a week on food and drink.' Sir Peter Phillips and Judge Griffith Williams also wrote in to complain, but after further discussions, an agreement was reached at £7,500.

There were also other issues for Michael Jones to deal with, including concerns being raised about the quality of food at some lunch sittings, as well as the inadequate costing of functions and some of the events which had been arranged to generate greater revenue. The international weekends had also seen overcrowding and had led to a series of complaints about the way the bar was being run.

But it was not all doom and gloom for the new Honorary Secretary as one of the positive spin-offs from the discussions with Oakley's, as well as the ideas floated for changes to the internal layout of the Clubhouse, was the need to have an independent audit of the facilities. Advice was taken

An Agenda for Change

from CADW over the best location for a lift and, by 2006, a full Disability Access Audit had been completed.

After receiving the findings from CADW and the Disability Audit, and considering the lift's impact, the committee agreed that the way forward was for a platform lift to be installed from the car park to the ground floor, with its location near the lavatory block, also requiring a re-arrangement of the washrooms and urinals. With this in mind, an additional sum of £30,000 was ring-fenced to cover these various works, plus any other items in order to comply with legislation. Tenders were sought and an estimate of £95,000 was duly accepted. Despite their delight at reaching a far cheaper solution than before, the committee realised that the whole issue of the lift installation had been very sensitive, and they informed the contractors that they would await approval at the AGM before allowing the work to commence.

This was achieved, and Michael Jones submitted a planning application for the work to begin. The initial application though was rejected, with a request that the lift shaft be encased in brick rather than being glazed. A revised application was approved in November 2008 and, after the necessary re-jig to the design, work began during the summer of 2009 on the installation of the lift, the creation of new washrooms, as well as a chairlift being installed to the upper floors. By December the work was complete and the various bits of paperwork associated with disability legislation were signed off.

The installation of the lift had been a major move forward but, by the time the lift was first used, there had been several important changes in personnel following the departure of Nicholas Cooke as General Manager on December 14th, 2007. His duties were assumed by Angela Williams, whilst Tim Warwick was invited by the committee to undertake a review of staffing and internal management procedures. Part of Warwick's review however created a number of additional problems, especially with the turnover of staff after several had been put on fixed-term contracts.

These changes in personnel, as well as Cooke's departure, almost inevitably caused plenty of comment and intrigue among members. But there was soon another topic for the members to mull over as news filtered through that Yates Wine Lodge, which had taken over the neighbouring Jackson Hall complex was likely to close because of financial problems caused by the downturn in trade. By July, the building was vacant and the talk among several members within the Clubhouse

John Cosslett.

The Westgate Messenger

THE NEWSLETTER OF THE CARDIFF & COUNTY CLUB　　　　　　　　　　　　　　　　　　　　DECEMBER, 2005

LEFT: The impressive façade of the Club, on Westgate Street

RIGHT: The rear elevation, which faces the Millennium Stadium

Mystery bidder hides behind a legal fence

Lawyers representing the man believed to be behind a £4.7 million bid for the Cardiff & County Club have now made a formal approach.

They have asked to meet the Club Committee in this connection, and are anxious to make a presentation detailing the would-be purchaser's plans for the building.

It is believed that there are no plans to demolish — or even alter the exterior — of the Club.

But a report in *The Western Mail* last week suggested that the building was one of three which might become the new official residence of the First Minister of the Welsh Assembly.

The others are the Judges' Lodgings at Radyr Chain and Insole Court, in Llandaff — both of which are richly-furnished Victorian mansions.

But the report quoted an Assembly insider as saying that the Club was likely to be the final choice because of its location in the middle of the city and its parking facilities.

It is known that Rhodri Morgan has been keen for some time on an Assembly building to compare with the city's Mansion House, which is used to house important guests and for civic functions.

There are also reports that the area of land between the Millennium Stadium and the rear of the Club has been snapped up in another secret deal by a firm of South Wales estate agents acting on behalf of an "official" buyer.

The price has not been made public, but is believed to have been in the region of £2 million.

Members of the Club have not been officially notified of the offer for the premises, but the Committee have given an assurance that as soon as they have enough information, they will summon an extra-ordinary general meeting to explain the proposals.

Many members are known to be aghast at the thought of selling the building, but a spokesman for the Committee made it clear that no action would be taken on the offer without a clear mandate from members.

One possibility that might arise is that new premises in a suburban location might be purchased with the sale proceeds.

A major benefit in that event would be that there would be an opportunity for increased parking facilities — and sufficient funds remaining to establish a sinking fund of sizeable proportions.

The Club's archivist, Brigadier Arthur Weston Evans, who is the author of the history of the Club, said that he would never agree to the sale of the building.

"We have inherited it from those members who went before us," he said. "They will surely be turning in their graves at the mere thought of the prospect.

"I am collecting names of members against the idea, and calling an emergency meeting."

He has the support of many of the legal fraternity who are members and who find the Club convenient both for its parking and dining facilities.

Silver hoard found in long-locked room

A recent visit to the Club by a team of surveyors led to the discovery of a locked strongroom behind shelving in the Manager's office. None of the present staff was aware of its existence.

Mr Cooke informed the Committee, who called in a locksmith. But he was unable to open it, and after a series of calls, Mr Cooke contacted the makers of the door — Laporte Frères, of Lille.

They were able to date the door to 1916, and turned up correspondence revealing that the Committee of the time were anxious to protect the Club's extensive silver collection during the Great War.

Laporte sent over Albert Ardeche, grandson of one of the team who installed the door, and with the help of the original blueprints, he was able to open it.

Inside were more than 60 pieces of table silver which are now being valued. The Committee will then decide what is to be done with them.

One of the 60 pieces of silver found in the room

French locksmith Albert Ardeche, who opened the door

An edition of a Club 'newsletter', published in December 2005.

was whether or not the Club would acquire the premises to create a gym or hotel accommodation, as had been mooted several years before, and raised again as part of the discussion stimulated by *An Agenda for Change*. Some members had suggested in the questionnaire that the Club could seek special arrangements with a nearby hotel whereby visitors and members could stay over at discounted rates, but now there was the possibility of the Club doing something itself and acquiring Jackson Hall.

But it wasn't just a group of members from the Club whose eyes were on the empty building as, for many years, the grandees within the Welsh Rugby Union had coveted the thought of having a hotel complex of their own adjacent to Westgate Street. In July 2008, they formally contacted the Club's committee to seek a meeting at which they could discuss acquiring the Clubhouse, or building over part of the Club's car park.

At their meeting in late August, the committee were adamant that the Clubhouse would not be sold to the Union, while little consideration was given to the proposal of exchanging part of the car park in return for increased parking provision elsewhere. The death of Sir Tasker meant that the club did not have a direct line into the inner sanctum of the Union and, during the autumn of 2008, there were some Club members who feared the worst as a meeting to discuss the Union's proposals was agreed upon for November. However, the downswing in the economy came to the Club's aid as shortly before the proposed meeting, the Union contacted the Club to cancel the get-together, stating that 'the credit crunch makes our scheme unviable.'

Within a few months, the Yates Wine Lodge had been converted into the WRU Shop and a base for their lucrative stadium tours. As visitors and shoppers flocked into the premises, the plans for a hotel complex and a so-called Westgate Plaza were shelved, much to the relief of the members of the Club.

30

Bleddyn and Dr. Jack

"The Club has been frequented by some quite remarkable men from such a broad cross-section of life in Cardiff."

In recent years, Cardiff has added significantly to its already impressive sporting c.v. by staging Test cricket as well as events in the 2012 London Olympics. As far as cricket is concerned, July 2009 saw Glamorgan's headquarters in the Welsh capital become the world's 100th Test Match venue as it played host to the opening match of the Ashes series between England and Australia. This followed the redevelopment of the Sophia Gardens ground, where Glamorgan CCC had played since 1967, and its metamorphosis during 2007 and 2008 into the stadium now known as the SSE SWALEC.

The opening match of the 2009 Ashes series – as six years later in 2015 – drew an estimated television audience in the region of 60 million, and each day, the 16,500 seater stadium was sold out as England and Australia went head-to-head. Given the high footfall expected through Westgate Street, the Club had wisely opted to offer breakfasts each morning for five days from July 8th, 2009, and again in 2015, to members who were visiting Cardiff on this most auspicious of sporting occasions.

Three years later more than a billion television viewers were tuned in as Cardiff became an Olympic city, staging a dozen matches between July 25th and August 10th, in the men's and women's football tournament. Indeed, the matches involving the women's teams from Great Britain, New Zealand, Brazil and Cameroon took place at the Millennium Stadium 48 hours before the glitzy opening ceremony in London itself. The quarter-finals of both competitions were in Cardiff and, on August 10th, the bronze medal game in the men's competition was held at Cardiff, when South Korea beat Japan 2-0.

Back in 2011, representatives from the London Olympic Games Organising Group had visited Cardiff, and specifically the Clubhouse, to review security arrangements and other matters relating to the football matches at the stadium. After taking the Group's advice, the Clubhouse was not fully operational during this two-and-a-half week period but, despite the fears of the Club's committee, there was not a significant loss of trade and the global sporting event did not have as adverse an effect on takings as some had feared.

Sadly, however, both of these showpiece international events were overshadowed by the death of two long-standing Club members and very prominent members of Cardiff's sporting community. For those attending the opening day of the 2009 Ashes Test, it proved to be a sad pilgrimage to the Club as two days before, Bleddyn Williams – one of the doyens of Welsh rugby during the 1950s – had died at the age of 86. For many years, Bleddyn had been a regular visitor to the Clubhouse, with staff welcoming him each Thursday as he dined and drank with a small group of friends, one of whom was Dr. Jack Matthews, another stalwart member and former Welsh rugby international, who passed away during the week before the start of the Olympic football matches in July 2012.

Back in the 1950s, Bleddyn and Jack were as renowned a double-act as Hollywood stars such as Dean Martin and Jerry Lewis, with the exploits of Cardiff's very own A-listers being woven into the annals of the city's famous rugby club. Known as 'the prince of centres', Bleddyn Williams won 22 Welsh caps between 1947 and 1955, besides boasting a 100% record of victories in the five matches when he led the Welsh side. One of the highlights of his distinguished playing career was the six-month-long tour with the 1950 British Lions to Australia and New Zealand – a journey he made with his Cardiff club colleague and life-long friend, Jack Matthews.

However, Bleddyn had been suffering from a serious knee ligament injury during the 1949-50 season, and it looked like he might miss the tour. The Lions selectors however insisted he prove his fitness after surgery by playing in a friendly at Bath – as Bleddyn gleefully recalled: "Cliff Morgan knew the score and we hatched a plan. Cliff made sure I wasn't overworked during the game but promised to make a break right at the end and 'gift' me a try to wake the selectors up. It worked perfectly, I am delighted to say."

Fortunately, Bleddyn was able to complete his rehabilitation on the six-week voyage to New Zealand and duly appeared in 20 of the 29 tour games, scoring a dozen tries and in the process garnering a host of tales of life on tour down under which he and Jack would readily recall during their get-togethers in the Clubhouse. 1953 had also seen Bleddyn enjoy the rare distinction of captaining both Cardiff and Wales to victory over the

The match programme from the 1953 clash between Cardiff and Australia, which starred Bleddyn Williams and Jack Matthews. It also included an advert (bottom right) from the company run by Club stalwart, Geevers, and his father-in-law.

All Blacks – a feat achieved within the space of three weeks, and toasted on countless occasions by his friends in the Club.

Born at Taffs Well on February 22nd, 1923, he played his part during the Second World War when, barely out of his teens, he trained with the RAF as a fighter pilot in Arizona before returning to the UK to learn how to fly gliders as preparations for Operation Overlord began. He flew a sortie with a cargo of medical and radio supplies to support troops involved in the Rhine Crossing. But he crash landed in an orchard, just inside enemy territory after being peppered by bullets from German guns.

He spent a week sleeping in ditches, with only his parachute to keep him warm, as he carefully made his way back towards Allied troops. When bumping into his commanding officer, Hugh Bartlett, DFC, who was a talented cricketer with Sussex, Bleddyn was afraid that he might ask the whereabouts of his glider. But Bartlett instead said to the young Welshman, "Williams, aren't you meant to be at Welford Road tomorrow playing for Great Britain against the Dominions? They need you. I think you ought to go now!" His response was in the affirmative, and a short while afterwards, Bleddyn caught the last supply plane of the day back to Brize Norton, before completing something of a *Boy's Own* tale by scoring the winning try in the match at Leicester.

This was another tale which Bleddyn would gleefully retell when in the company of Dr. Jack, whom he first met in 1942 while playing for the South Wales Anti-Aircraft XV. They formed a fine combination for Cardiff RFC, especially when carrying out a series of special scissor moves. "I always started with the ball and he always would cut the angle," Bleddyn recalled. "On the first occasion, I would feed Jack and he would get tackled. The second time I would feed him and he would get roughed up again, but he would always get up and wink at me, saying 'You know what to do now, Bledd'. So on the third time, I would throw the dummy. He would get hit by two or three hard cases, but I would saunter off to score under the posts and get all the glory and headlines – but they were Jack's tries really."

Bleddyn Williams.

The Medics

Dr. Jack Matthews was one of the high profile members of the local medical community to be a Club member. Others in recent years have included Professor John Lazarus, Dr. Peter Beck, the late Iain Breckenridge, John Stradling Griffiths and David Watson James, but the percentage of doctors and dentists who are members of the Club is somewhat lower than other professions.

Professor Lazarus is one of a group of medics who dine in the Club on Wednesdays. He joined the Club in 2001, and in subsequent years has played a key role by serving a total of six years on committee which included a significant and important spell as Chairman of the House Sub-Committee. Amongst the projects the Scottish-born medic has been closely involved with were the establishment of a lift to the upper floors, and the refurbishment of the Jubilee Suite, besides keeping a watchful eye on the maintenance of the Clubhouse's external fabric, especially the gable end above the Westgate Suite which had been in danger of collapsing. Educated at Rugby School, Queen's College, Cambridge and Glasgow University, John is one of the country's leading experts on clinical endocrinology and the functions of the thyroid, and it is through his work and friendship with Dr. Peter Beck at Llandough Hospital that he joined the Club.

Since joining in 1989, David Watson James has given sterling service to the Club, serving as Chairman on two occasions and helping the organisation through some difficult years when financial worries and staffing issues proved challenging. Educated initially at Marlborough Road and Roath Park Elementary School, David attended Llandaff Cathedral School and Malvern College. While at the latter, he was fortunate enough to be in the school's Combined Cadet Force which formed the Guard of Honour at the Victoria Memorial opposite Buckingham Palace during the Queen's Coronation in 1953.

After reading dentistry at Birmingham University, he worked briefly in the Midlands city as well as Great Malvern before buying a practice in Cathedral Road in 1961. During his time as a dentist in Cardiff, David

David Watson James (far right) with Nigel Phillips, Judy Watson James and Agnes Phillips.

became involved with various national organisations including the General Dentistry Services Council, of which he was National Chairman in 1982. He served for five years on the General Dental Council besides advising Denplan on many aspects regarding insurance. He also worked for the Royal College of Surgeons and helped to set up the Faculty of General Dental Practitioners.

In 1998 David joined the committee of the Club, and two years later took over as Chairman in the wake of the spat between Sir Geoffrey Inkin and Sir Tasker Watkins. As David recalled: "Geoffrey had initially said at the committee meeting following the AGM that he would be prepared to carry on, but on his journey home he had second thoughts and a couple of days later John Payn, the Honorary Secretary, asked all of the committee to attend a meeting at the Club."

"Everyone was quite shocked at Geoffrey's decision so it was agreed that Paddy Gallagher would take soundings. My name was the one that came forward so I agreed to take over. Given the heated discussion at the AGM, and the potential for further disputes, I realised that there was a need for consistency so at the first meeting as Chairman, I had a motion passed that there would be no changes in the rules whatsoever during the course of the coming year. This succeeded in calming things down."

David's deft hand was evident again during his second spell as Chairman at a time when there was a need for refurbishment of the dining facilities as well as the creation of a place where ladies could dine. After hearing from Keith James about developments in this direction at the Oxford and Cambridge Club, David turned his and the committee's attention to what was the committee room at the front of the building. In David's own words, it was "similar to a Welsh crematorium on a wet Sunday," so a more tastefully mixed dining area, known now as The Westgate Room, was the outcome of his efforts.

Bleddyn and Jack remained close friends for the rest of their lives, with Jack joining the Club in 1972, four years before Bleddyn, who was the rugby expert with the *Sunday People* newspaper for more than 30 years. The pair would spend Thursday lunchtimes in the dining room before spending several hours in the company of several of the other larger-than-life characters in the Club, including Howard Jackson, a charming businessman with a great love of rugby, and Paddy Gallagher, an Irishman who was prominent in the local building trade.

Bleddyn also sought out Jack's company at Wales' home games at the Arms Park, with Bleddyn in his Donegal tweed jacket sitting quietly with Dr. Jack and reminiscing about their own playing days. When the New Zealander Graham Henry became the Wales rugby coach in 1998, one of his first acts after arriving in the country was to seek out the pair; his father had watched the two centres during the Lions tour to New Zealand in 1950 and never tired of telling his son how highly he rated them.

The short, squat figure of Dr. Jack was once described as being 'a cross between a bulldozer and a brick wall.' Nothing got through him and his strength with the ball in hand created many holes for his friend to run through. Indeed, when Bleddyn retired in the mid-1950s, his tally of 185 tries was a club record for Cardiff. Regarded as the iron man of Welsh rugby, Dr. Jack also fought a four-round draw with Rocky Marciano when the future world heavyweight boxing champion was stationed with American troops in Wales during the Second World War.

A GP in Cardiff for much of his life, Dr. Jack claimed to have brought some 7,500 babies into the world, and sometimes worked for many long hours during the night before an important game of rugby. He won 17 caps for Wales between 1947 and 1951, and led the country to victory in what proved to be his final international appearance against France in Paris. His international career, however, ended rather abruptly shortly afterwards, when he tore a strip off the selectors in very salty language after they had summoned him from Cardiff on an overnight train as a late replacement, before telling him just ten minutes before kick-off that his services were not required after all. He remained a man of firm character and as fearless in his opinions off the field as he was effective on the field with ball in hand alongside Bleddyn.

Born in June 1920 at Bridgend, he attended the Welsh National School of Medicine, and as a teenager, won the Welsh AAA junior 220 yards title. Like Bleddyn he also joined the RAF, but after the military authorities discovered he had embarked on his medical studies he was switched to the Royal Army Medical Corps. He said: "I was disappointed not to fly, but 85% of my pilot intake did not survive the war. I was truly one of the lucky ones. Our generation — the ones who survived or were spared — have spent the rest of our lives trying to make every last minute count one way or another. It was the deal we struck privately with ourselves to keep our sanity and to honour those who didn't make it."

As one long-standing Club member said: "The death of Dr. Jack marked the end of an era for the Club. He was amongst the last of a group of members who survived the Second World War besides forging a hugely successful sporting career with a life in other areas of what makes Cardiff tick, and all after he, like his great friend, Bleddyn, had fought in the Second World War when barely out of school. We have so much for which to thank their generation."

These heartfelt sentiments were shared by so many other members and there had been a feel-good factor within the Club in March 2006 when the committee agreed to a joint portrait being commissioned of the two former Welsh rugby internationals. It now hangs above the staircase leading up to the first floor, and is adjacent to the painting of their other old pal Sir Tasker, which had been commissioned three years before.

This has been just one of many changes to the internal fabric of the Clubhouse in the past few years, although some of these changes have been prompted by other more sinister events. For example, there was an attempted break-in via the reading room during February 2009, and the upshot of these unsuccessful efforts was the installation later in the year of intruder alarms. During 2010 new carpets and furniture were bought for the hall, stairs, bar and dining room, and humorous cartoons by the late Austin Thomas and other amusing artwork were relocated in the washroom area.

The Club's virtual presence was revamped following sterling work by Jon Wooller, Howard Wilkins, Ceri Preece and Philip Azzopardi. Their efforts led to a revamped website being launched in November 2009 with full details about the facilities, the splendid food and beverages, membership and the ever-increasing range of special interest groups. Indeed, by this time, the Club also boasted a Cricket Society and a Literary Circle, as well as the long-standing Golf Society which continued to go from strength to strength with annual events at the Royal Porthcawl links.

These special interest groups have significantly added to the kudos of the Club, and a host of top-flight speakers, as well as leading personalities in the worlds of literature, the arts, religion and sport have spoken in recent times at the Club. They included Dr. David Jenkins of the National Museum of Wales; The Reverend Doctor Barry Morgan, the Archbishop of Wales; the late Professor John Davies, who was an expert on the history of Cardiff and the Bute Estate; David Cornock, BBC Wales' political correspondent; Gwyneth Lewis, the former National Poet of Wales; Baron Morris of Aberavon; Carolyn Hitt the journalist and broadcaster and Owain Arwel Hughes, the famed conductor.

Roger Gagg, the former chairman of the Friends of the National Museum of Wales, has been a leading figure with the Literary Circle. An accountant by profession, Roger has been a member of the Club since 1983, and in addition to acting as Chairman of the Literary Circle, he has chaired the Wine and Social Committee, whilst spending three years on the General Committee. Created during the tenure of Michael Jones as the Club's

The piping-in of the haggis at the Burns Nicht event in the Clubhouse, January 2016.

Hononary Secretary, the Literary Circle meet for lunches six times a year, and in Roger's own words "The aim of the meetings is to hear a speaker who will appeal to a broad spectrum of the membership, thereby adding a cultural element to the rich diet of events in the Club."

A hugely successful eve of Test Match dinner in 2011 when England met Sri Lanka at Glamorgan's headquarters, saw a range of cricketing names in attendance. Other cricket luminaries to speak at the Club have included: Hugh Morris; Graeme Swann; Phil Tufnell; David Collier, the CEO of the England and Wales Cricket Board; Ian Chappell, the former Australia captain; Glamorgan legends Robert Croft and Mark Wallace; selector Geoff Miller; and former England batsman Mike Gatting. The Cricket Society have also organised outings, with the group visiting the first day of Glamorgan's County Championship match in 2010 against Gloucestershire, which was held at Cheltenham as part of the popular Festival Week of cricket, when members from the Cardiff and County Club joined members of the New Club in their marquee at the delightful Cheltenham College ground.

The St. David's Day Dinners, as well as *Burns Nicht*, have become hugely popular, together with a host of other evening events such as Gourmet Evenings, Guest Chef Nights, an Indian Banquet, a West End Musicals Evening, a Shellfish Spectacular, Jazz Sunday Lunches, evenings focusing on Burgundy, Bordeaux and New Zealand wines, as well as tutored wine-tasting sessions. Breakfasts have also been introduced on weekday mornings from 8am, adding further to the rich diet of Club life, together from February 2016, with a lighter lunch menu and afternoon teas. These ideas for evening events, especially the Guest Chef nights stemmed from the tenure of Alex Embiricos as Chairman from 2012-13 to 2013-14. Born in Cardiff to a well-known shipping family, Alex remains one of south Wales' leading tax experts and besides having been partner in KPMG, he has served on the council of Cardiff University. He has been a member of the Club since 1992 and has been delighted to witness the success of these evening events, both in terms of their social function and their profitability, besides the way they have added more adventurous and innovative cuisine to the Club's offerings.

The evening events have helped to promote the benefits of membership, proudly shown off from 2012 at a prospective members' evening. A number of other initiatives have been considered in the past few years, with research being undertaken by Alun Davies, who chaired a membership working party, into the membership arrangements at other clubs in the UK. The research showed that the Cardiff and County Club was among the most expensive in the country and, overall, had the highest joining fee. This prompted a call to drop the joining fee to £200 and abolishing it altogether for those aged under 30. Michael Jones also undertook a lot of work fostering links with reciprocal clubs, but the suggestion that incentivising members to propose new candidates was, quite rightly, deemed as being inappropriate for such an organisation.

Getting people through the mahogany door and into the wood-panel lined lobby is still seen as being a vital ingredient of recruiting new members and, with this in mind, the Club is still delighted to be hosting gatherings of Old Boys from Monmouth School, Clifton College, Christ College Brecon and Llandovery College, as well as functions held by various Rotary and Probus Clubs, the Cardiff 41 Club, and groups from the legal world and other professions.

The fact that, year after year, these organisations visit the Club is testament to the high standard of food and wine on offer, something which the present members of the Wine and Social Committee are at great pains to maintain. At a time when there are many other eateries in Cardiff, as well as gastropubs and themed restaurants, the view is that the Club still remains the finest place in Cardiff to dine at lunchtime. With recent changes in the membership and other future initiatives, it could yet become the place to eat in the evenings, right in the heart of the Welsh capital city.

31

It's a Man's World No Longer

"It's a place so loved by so many people. I cannot imagine what Cardiff would have been like without it."

"It's without doubt the most significant change in the entire history of the Cardiff and County Club" – the words of long-standing member and former Chairman, David Watson James, following the decision taken during 2014 to admit lady members. The question of ditching male-only membership had been rumbling on for many years but, by the second decade of the 21st century, there was a growing consensus that the time was right for a sea change in membership.

It's impossible to pinpoint a single reason or event prompting this change; a number of underlying factors played a contributory role. Some were related to changes in society with an increasing number of important organisations and their top executives regarding the Club as a sexist organisation. Ironically, by the second decade of the 21st century, there were fewer misogynistic views being aired behind the mahogany door of the

Prof. John Lazarus (left), Ceri Preece and their wives, Maureen and Helen (right) at a function in the Clubhouse.

Clubhouse, but the male-only rule was a tangible sign to many potential members that the Club was stuffy, old-fashioned and out-of-touch with modern life.

There were economic factors as well, not least the need to maintain the Club's revenue at a time of economic hardship, as well as legislative ones, following the passing of the Equality Act in 2010. It came into force on October 1st, 2010 and required all guests of the Club to be treated equally. The discussions within the committee room therefore set the tone, with the conversations relating to compliance with the Act being the catalyst behind a series of changes leading up to the ground-breaking decision at the 2014 AGM.

It did not take the committee long to agree that, in order to comply with the Act, from March 1st, 2012 all guests of the Club should have access to the bar as well as the main dining room. However, there were many other issues to consider, not least revisiting the question of admitting lady members. Previous experience had highlighted the need for careful consultation with members as well as transparency in all of the discussions, so a Membership Sub-Committee was formed, chaired by Paul Twamley, a former Chairman and Club Trustee, and including Alex Embiricos, Professor John Lazarus, Harry Lewis and Andrew Williams, to consider the issues relating to the Equality Act, the Club's Rules and Bye-laws, as well as other matters relating to membership of the Club.

The opinion of Jonathan Walters, a senior counsel specialising in discrimination law, was sought as well as those of the Club's past and present Solicitors, Lawrie Williams and Adrian Larby. As a result, a consultation document was prepared ahead of a briefing meeting with members on March 28th, 2012. The key issue behind the meeting was to get members to consider all of the requirements of the new Act, rather than just focusing on the issue of admitting women members. Hence the consultation

Alex Embiricos (left) with John Weber, Lizzie Embiricos and Nigel Phillips.

A function in the Westgate Room.

document looked at other broader issues relating to membership, including the use of the rooms and the dress code, with the following ten key questions being raised:

1. Should the Club revert to a members-only club or have a male-only guest rule and a male-only associate rule?
2. If members wish to invite guests to the Club, should this be limited to specific days of the week, certain times of the day or a combination of both?
3. If guests are to be allowed into the Club should certain areas be set aside for the exclusive use of members?
4. Should a charge be made for guests, either as an entrance fee or as a supplement to lunch or bar bills?
5. Should the Club open membership to ladies?
6. Should the number of times a guest is permitted to attend the Club be changed?
7. Should private functions at the Club be treated differently from Club functions?

8. Can the space in the Club be used more effectively?
9. Is the current dress code an inhibiting factor in your use of the Club?
10. What changes to the Club rules or to the physical layout of the Club would make you feel that your subscription offered more value for money when compared with guests?

Questions eight and nine touched on other sensitive issues, not least the fact that a significant sum was still being spent on maintenance, and an increase in income – either through more members or more functions – would be most welcome. The Westgate Room, where mixed dining took place, was being used occasionally as a function room, and the card room was being kept in reserve for mixed dining when the Westgate Room was in use for a function. The upstairs rooms and the Jubilee Suite were hardly ever used, despite the fact that successive committees had considered how they could be better utilised. All of the suggestions had fallen foul of the lack of money for renovation, and the fact that the expenditure was unlikely to be swiftly covered by an increase in income.

By getting the membership to think of the big picture, and the Club's future, the consultation document placed the question of lady members in terms of the changing socio-economic environment of Cardiff in the enlightened world of the 21st century. After reading the document, few members could have failed to make the link that allowing women to join would boost revenue, but the committee was adamant that no voting should take place on any of the issues, and that instead there should be a series of rational, civilised and informed discussions on all of these issues.

Indeed, had a vote on question five been taken over the admittance of ladies and the motion been defeated, it would have swiftly led to much publicity in the media besides creating an adverse – and incorrect – view of the Club and its members, especially at a time when there appeared to be a growing mood for change to the membership. This was clearly evident from the results of a questionnaire which was also issued during 2013 to canvas wider opinion from the entire membership. The overall result was 71-61 against admitting women, but there were many impassioned responses to the questionnaire in favour of the proposal. Out of the many comments, perhaps it was the following three which resonated most and confirmed that the time had come to admit lady members:

'The Club has to accept this is the twenty-first century. We should move with the times. Lady Members would enhance the Club and, of course, it makes financial sense.'

'Admitting Lady Members would revitalise a Club that is seen by many as irrelevant to today's society.'

'Top people do not feel that they can seek membership of a sexist organisation.'

The questionnaire, briefing papers and consultation meetings with members had given the committee the information it required. So after further discussion, it agreed at its meeting on August 15th, 2013 that it would recommend to the Club's AGM in 2014 that the rules of the Club be amended so as to permit the introduction of ladies as members of the Club. Given the fact that the majority of responses to the questionnaire had been in favour of retaining the *status quo*, the committee agreed that Julian Rosser, the Chairman of the Membership Sub-Committee, should send a letter to all members outlining the rationale behind the committee's decision to admit lady members. His letter read as follows:

'We are lagging behind the field…as even our most local reciprocals, such as the Bath and County Club, The Clifton Club (Bristol) and the New Club (Cheltenham) admit female members.
　The importance of this proposed change in our constitution is not just to conform with more enlightened thinking (and we do not anticipate a sudden surge of female applicants), but to allow various organisations, including political bodies and NGO's, which currently do not approve of male employees joining what is perceived as a sexist organisation, to change their view.
　We believe that, maybe somewhat perversely, we will enjoy an increase in male applicants as a result of what we see as collaborating with the inevitable.'

The 2014 AGM agreed broadly with the sentiments of the proposal but, given the results of the questionnaire, it was agreed that further discussion would take place, with a briefing meeting being held in October 2014 to ask members whether or not they would be in favour of holding a vote on women members. They agreed and a Special Meeting was duly held on November 25th, 2014 when a gathering of some 175 members took place in the Clubhouse; a 75% vote in favour of the proposal would be required for the resolution to be passed.
　Professor Dylan Jones-Evans, a well-respected and relatively younger member of the Club, had expressed the feelings of many by writing in his column in the *Western Mail* that 'it is a complete anachronism that

an organisation that is an integral part of the business community in the capital city of Wales does not allow women members.' It came as no surprise therefore that, unlike previous meetings when the question of lady members had been raised, nobody was prepared to stand and speak against the proposal.

The litmus paper had clearly changed, and the forward-thinking committee boldly nailed their colours to the mast of the reformers rather than the traditionalists. As it turned out, there was almost an 85% majority with 146 voting in favour of the motion of admitting ladies, and just 24 people voting against. Within a few minutes of the decision being ratified, the news that women members would be accepted appeared on social media – evidence if any were needed of the changing world in which the Club found itself. The following morning, stories were carried on BBC Radio Wales and in the *Western Mail* with the newspaper carrying the following comment from long-standing member Alun Davies – 'The Club was formed when men ran businesses, there were no women in offices, 50% of fee-earners in firms are now women. The times are changing.'

The First Lady Member

In 2015 Fiona Peel, a very well-known face in the world of health in south Wales and beyond and former High Sheriff of South Glamorgan, was the Club's first female member.

Educated at Moreton Hall in Shropshire, and St. Thomas' Hospital in London, Fiona worked as a nurse before getting married and moving to Devon where she read history at Exeter University. In 1979 she and her husband moved to south Wales to set up the Berry Hill Fruit Farm near Castleton, before Fiona began her distinguished career in public service.

After serving on the Board of Governors at Bassaleg School, she was a non-executive director of the Gwent Health Authority before being appointed to the chair of the Gwent Community and Mental Trust. Fiona also took a postgraduate degree at Cardiff University in the legal aspects of medical practice. She served on the council of the

Fiona Peel.

General Medical Council for six years until 2009, and was chair of Cardiff Local Health Board from April 2006 to March 2009 when all LHBs were re-organised. At present she is on the council of the General Optical Council.

Awarded an OBE in 2001, Fiona was appointed as a Deputy Lieutenant of Gwent in 2003, before serving the following year as High Sheriff of South Glamorgan. She has also been chair of the Merger Commission for Cardiff University and the University of Wales College of Medicine, as well as chair of the Cancer Services Co-ordinating Group in Wales and the Cardiac Network Co-ordinating Group between 2002 and 2010. Since 2011 she has been a trustee of the Tenovus cancer charity.

As she freely admits: "I had always wanted to be a member because I believe in traditional values. The Club is very good as it is – the only thing that it just needed to accept was equality. In my roles in public service as well as serving as Deputy Lieutenant and High Sheriff being unable to become a member of the Cardiff and County Club became something of an irritant.

"As far as so many members of the Club are concerned, I'm not really an alien face. So many were already friends or have worked with

H.R.H. The Duke of Edinburgh meets Michelle Tuck, the Club's first female committee member. Also in the photograph are John Weber and Ken Truman.

me in public service for many years. I usually come in for lunch on a Friday, and feel welcomed at practically any table. Everyone has been so positive, but I'm not a revolutionary and I have not come in wanting to change things or make a fuss. Of course, over time we must repair and revive just as we do in all of our personal lives.

"I like and appreciate everything that the Club currently offers and intend ensuring that this remains, besides supporting other initiatives as the Club moves forward. By the time the Club celebrates its 175th anniversary, I'm sure that at least one woman will be on the committee, but in a world of equality why shouldn't there be!"

Indeed, Fiona's closing comment became reality in June 2016 with the election of Michelle Tuck as the first female committee member.

A meeting in the Club's committee room.

32

Onwards and Upwards

"The Club has adapted to change, and is now prosperous and successful, with a stable membership."

One of the enduring sights greeting members and visitors alike in the lobby area of the Clubhouse is the rack of newspapers and, in the reading room, other periodicals and reading material. Over the years, the front pages of these publications have carried headlines and stories about kings, queens, coronations and wars. Here, within the welcoming walls of the Clubhouse, members have been able to catch up quickly with the news of the day or, at leisure, delve into the broadsheets and read in more detail about events of local, national and global significance.

With talk of business frowned upon in the main dining room, it has been the newspapers and the other publications which over the years have been the source of either comforting or troubling news about the state of the local, and British economy. Details of the Wall Street Crash of 1929, the trade depression of the early 1930s, and in more recent times, the so-called Black Monday of October 1987, and Black Wednesday of September 1992, as well as the credit crunch of 2008 and 2009 have all been devoured and dissected by members.

The Club itself could have had a double-dip recession of its own, following the

Owen Sennitt.

departure – within the space of 12 months in 2007-08 – of General Manager Nicholas Cooke and Honorary Secretary Michael Jones. The fact that both left the Club within a short space of time could have had a destabilising effect. Peter Dewey had also retired as Honorary Treasurer in October 2007, and Head Chef Mark Jones left in October 2009. But through strong leadership, via various committee personnel and the appointment of able replacements, the Club has moved onwards and upwards, and avoided some potentially difficult times.

Andrew Williams (far left) with Chris Clarke and Hefin Owen.

A number of amendments to working practices and personnel have also taken place during the past seven years, but staff morale within the Club is now at its highest for many years. The likes of Paul Twamley, Jeff MacWilkinson, Professor John Lazarus, Bob Edwards, Owen Sennitt, Alex Embiricos, David Mansel Jones and David Watson James have all played significant roles through serving on the committee, together with the affable Honorary Treasurer, Andrew Williams.

Born in Rhiwbina, Andrew was educated at Llandaff Cathedral School and Christ College, Brecon before reading Electronic Engineering at Liverpool University, and then training to be a Chartered Accountant. During his career, he has worked for companies in Aberdeen and Arizona, before returning to his native Cardiff.

As befits the son of Gwyn Williams, a prominent cricketer with Cardiff CC, Andrew is a well-known figure on the local cricket scene, besides managing the Glamorgan Greencaps, the team for players with disabilities, who train at the county's facilities at The SSE SWALEC and play fixtures against other teams from English counties.

For Andrew, the Club is a very special place: "It is a place to meet people, discuss things and hear the views of the opinion formers of Cardiff. At its core are the quality of people and standards, and in the heart of a busy city it's an oasis of calm and tranquility."

Honorary Secretary Ceri Preece skilfully guided the Club up to, and through, its 150th year. Ceri Preece, a well-known local solicitor, was

Ceri Preece with his wife, Helen (right), and Mrs. Rhian Sennitt (left).

formally appointed the Club's Honorary Secretary following the sudden departure of Michael Jones. Born in Port Talbot in 1955 and educated at Ysgol Gyfun Rhydfelen and Aberystwyth University, Ceri has worked for Hugh James Solicitors since 1978, specialising in property law. A Club member since 1994, and a former President of Cardiff Chamber of Commerce, he swiftly got his feet under the table in his new role at the Club. Having taken over on February 2nd, 2009, he immediately had to deal with the issuing of annual subscriptions as well as a request for a Special Meeting as some members felt unhappy at the way in which the services of some staff had been dispensed with some years previously.

Among his other tasks was correspondence with Cardiff City Council about the maintenance of the bus shelters which, in some people's minds, had almost been put in their precise locations to cause the greatest amount of disruption to the Club and its members. Together with Jonathan Arter, Lawrie Williams and Graham Walters, Ceri also helped to undertake an overhaul of the Club's rules relating to disciplinary procedures, to unpaid subscriptions, and to members' access to minutes, as well as bringing in new working practices such as allowing the silent use of electronic devices throughout the Club.

Following the resignation of Angela Williams in April 2012, Ceri also undertook the additional duties of acting General Manager for a couple of years while conversations took place,

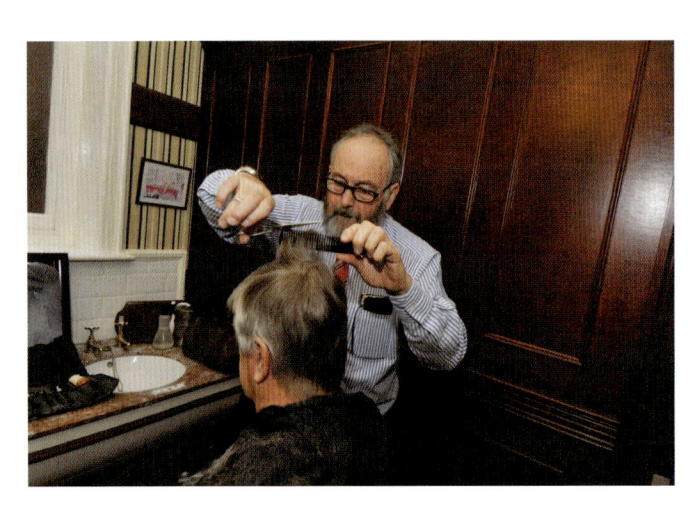

Mr. Glenvin Prisk, the Club's Barber.

at committee, about the administrative structure of the Club and its needs in the modern world of email and cyberspace. One of the challenges he successfully tackled related to the catering operations on the days when the Welsh rugby team are playing at the Millennium Stadium, with two serving times being introduced in the dining room, and the installation of a secondary bar upstairs together with a big screen showing the television coverage. One of the more unusual initiatives he has also overseen has been the installation, since February 2013, of a barber shop in the gentlemen's washroom, with Glenvin Prisk available by prior appointment between 12 noon and 3pm on Tuesdays, Thursdays and Fridays.

September 2014 saw Essex-born Nick Lawrance join the Club as General Manager. He had a wealth of experience in the hotel industry, especially relating to food and beverages, having worked initially for the Marriott and Holiday Inn chains in both Plymouth and Portsmouth before moving to Cardiff to become general manager of The Angel Hotel. After several years

Nick Lawrance (far right) with Club staff including Head Chef Lee Thomas.

at the Angel, Nick became a consultant in hotel management and worked in various four and five star hotels in Sydney, Australia as well as at the Future Inn in Cardiff and then the Ashton Court Hotel in Bristol, before joining the Cardiff and County Club.

Besides admitting women as members, the decision-makers within the Club have made significant changes to the dress code – a matter which had become a constant sore point for some members. Attempts had previously been made to relax the code at weekends and on international rugby days, although the latter had led to confusion over whether the wearing of rugby shirts, jeans and casual training shoes was acceptable. While some were happy to relax the need, at certain times, for ties or jackets, others argued that a relaxed standard of dress could lead to a lower standard of behaviour and other problems. After careful consideration of these issues, as well as feedback from a members' questionnaire, a decision was taken to relax the rule about the wearing of ties on Friday. From September 1st, 2013 these were no longer required to be worn, although most members still opt to wear a tie when dining on Fridays.

This includes the group of long-serving members, known as the 'Debs', who presently comprise Lawrie Williams, Chris Brain, Michael Clay, John Ingledew and others who are the descendants of some of the Club's oldest families and whose forefathers were prominent in the hunting, sporting and business circles of south Wales. Their regular presence, at a table at the far end of the main dining room and adjacent to the balcony windows, is a reassuring reminder of the past, especially at a time when many other traces of the past have disappeared from around Cardiff, not least in Cardiff Bay where the Clay family made their fortune, or in the Brewery Quarter of St. Mary Street where a host of restaurants, pubs and wine bars now replace the buildings which were run so successfully by the members of the Brain family for many years.

Another key figure in the recent life and decision-making of the Club has been Paul Twamley, now a Club Trustee and the current Vice-President. Paul has been a member since 1981, and enjoyed a highly successful career with Deloitte before becoming group finance director of Hyder PLC and subsequently chairman of the Harris Pye Group who, among other local projects, has restored the Penarth Pier and Pavilion. He has been Chairman of the Club, besides being the key figure

Paul Twamley.

on the Membership Sub-Committee that advised on aspects of the 2010 Equality Act and the introduction of women members.

He had long been an advocate of women members and, following the rule change in November 2014, Paul is supremely confident that the Club will still be an important part of Cardiff's social, economic and political worlds in the coming years. "Admitting ladies was simply the right decision and one which will allow us to move forward," he says. "The increased subscriptions from both men and women will help to stimulate further growth and demand for our ever-increasing portfolio of events, besides generating further capital for maintenance of our premises as they get older and older."

Owen Sennitt was the Club's Chairman during its 150th anniversary. Born in London to a Welsh mother, Owen read Economics and Accountancy at Loughborough University before working in Cardiff from the mid-1980s, and becoming a specialist in corporate finance. Owen has also worked in London and Bristol with various venture capitalists, besides being involved in a number of corporate buy-outs.

A regular visitor to the Club, with business colleagues and acquaintances from the financial world, Owen joined in 2004 and was elected to the committee in 2013. Within a year of his election Owen was appointed Chairman and, at 44 years of age, became one of the youngest members to serve in this position. During his time in the Chair, Owen oversaw the process by which a new General Manager was appointed, served on the 150th Anniversary Committee and involved himself with discussions to move the Club forward whilst retaining its character and charm.

How Costs Have Changed

It is extremely difficult to compare the prices of 1866 with those of 2016 but, with some clever mathematics, it has been possible to provide the following comparisons between the start and finish of the Club's first 150 years as well as showing where expenditure has changed.

Other comparisons between the published accounts of 1866 and 2016 illustrate how costs for other aspects of the Club's operations have also changed:

	1866	2016
Printing and Stationery	£25.70	£9,122.77
Postage	£11.62	£ 602.71
Bank Charges	£ 1.20	£6,340.85
Clock Repairs	£ 5.75	£1,105.00
Newspapers	£50.22	£3,269.00

As far as the bank charges in 1866 were concerned, these relate to the costs of cheque books. The modern figure reflects the high use of credit cards and the charges they incur.

Indeed, the upkeep of a Clubhouse built in 1892, has been one of the chief concerns for Owen, his fellow committee members as well as the officers and staff at the Club. Back in 2010, Jon Wooller and others on the House Committee had been optimistic of obtaining grants from CADW to cover up to 30% of the costs of maintaining the fabric of the building. But the organisation replied that the building did not reach the very high standard required under the Historic Buildings and Ancient Monuments Act of 1953. It was unfortunate that, in 2011, parts of the basement and ground floor were flooded and, since 2012, the committee has worked closely with a number of advisers to ensure that there is no repetition of this, besides seeking a viable solution to the problem of damp in the Jubilee Suite.

The external appearance of the Clubhouse has continued to be a matter of concern, and remedial work has been undertaken on exposed patches of sandstone. Attention has also been given to the roof and chimneys to ensure that they are both wind and water-tight. All of this, however, requires both a methodical programme of maintenance as well as sufficient funds to allow this work to continue and to preserve the fine standing of the building within Wales' capital city.

But as well as keeping a close watch on the building and the inner fabric of the Club, an eye will also need to be kept on the resurrection of the proposals of the WRU for a hotel complex adjacent to Westgate Street.

As well as threats, there are also several positives, particularly with the 'Central Square' redevelopment on the site of the old bus station and its surrounding buildings between the railway station and the Principality Stadium. With many established businesses seeking to relocate to the new development, as well as the new BBC Wales Broadcasting House, there is great scope to attract current and new members to dine more regularly at the Club. There is also great potential to utilise the Clubhouse's upper floors for office space and 'hot desks' for members looking for a working base or meeting area in the heart of a vibrant city.

150th Anniversary Celebrations

The Club's 150th anniversary was duly celebrated in some style during 2016 by its members and their guests. A programme of special events was approved by the Anniversary Sub-Committee, Chaired by Clive Johnson, with the highlights including a Gala Dinner on Saturday June 4th in the Grand Hall of the National Museum of Wales in Cathays Park – another building with close links with the Marquess of Bute, on whose land the civic centre was laid out in the early 1900s. The evening, attended by 264 members and guests, saw a superb meal, planned and overseen by Stephen Terry of The Hardwick near Abergavenny, with the gastronomic delights being supplemented by a wonderful series of renditions by the internationally acclaimed soprano, Rebecca Evans, and the world famous baritone, Sir Bryn Terfel.

Rebecca Evans and Sir Bryn Terfel perform at the Gala Dinner.

Another major highlight was a visit in March by H.R.H. The Duke of Edinburgh, the Club's Patron, who joined 100 members for a special lunch and reception held in the Clubhouse. Masterminded by Ceri Preece, the event saw Prince Philip make a short speech before toasting the Club and then meeting all present, including the staff who helped to ensure that this royal visit went without a hitch and was a fitting way to crown the Club's 150th anniversary year.

> At the Club's AGM on June 29th, Clive Johnson was elected as the Club's next Chairman. Educated at Canton High School and Llandaff Technical College, Clive is a retired engineer who has spent the majority of his working life in the Welsh capital and has been a Club member since 2001. He and his wife Sally have two children, seven grandchildren and a golden retriever which he freely admits demands more time than all of the others. However, as the anniversary events of 2016 have shown, Clive is an energetic gentleman and as befits a keen sailor, he will be at the helm as the Club continues its voyage towards its 175th anniversary.

The Cardiff of 2016 is very different from the relatively small town of 1866, when the Cardiff and County Club was established. Modern-day

It was not just the Cardiff and County Club who were celebrating in 2016. The successful Welsh football team pass the Clubhouse along a packed Westgate Street on their open-top parade through Cardiff on July 28th, following their inspirational performance at UEFA's Euro 2016.

Cardiff is home to 350,000 people and is growing fast, with projections for an additional 100,000 inhabitants over the coming decades.

As with most of the old county of Glamorgan, the city's economy has changed beyond recognition, with the export of coal and iron-ore being replaced by a range of new manufacturing and service industries as well as buoyant educational and tourism sectors. These have seen new and significant imports from all over the world in the shape of students, tourists and sports fans who generate around £1 billion to Cardiff's economy and maintains the cosmopolitan essence of Cardiff which, when the Club was founded, had been the preserve of Tiger Bay and the dockland communities that surrounded it.

Yet throughout this period of change, the Club has been a constant – maintaining high standards of conviviality and fellowship, with the great and the good of the city unwinding, initially in the premises at, and adjacent to, the Royal Hotel, before moving to the Clubhouse in Westgate Street. There may be fewer shipbrokers and grandees from the world of coal on the Members' Roll, but 'movers and shakers' of modern Cardiff can still be found enjoying the Club's good food, excellent wine and entertaining company of other members. As the Cardiff and County Club moves towards its 200th anniversary, these traditions of friendship and *bonhomie* will be maintained as the men and women of the Welsh capital city enjoy all that is good of being part of Wales' premier private members' club, and "always amongst friends".

Index

Alexander Family (of Gileston) 15, 134-142
Allen, Walter 111
Arter, Jonathan 280
Asscher, William 242
Azzopardi, Philip 267

Barlow, Tom 126
Bassett, Richard 17, 87
Beck, Peter 263
Berkeley, Charles 25, 29, 59-60, 69-71, 77, 104, 111
Bernard, Charles 15
Berry, Anthony 202-203
Biggs, Jacob 34
Bird, John 16
Box, Donald 213
Brain Family x, 61-63, 65-66, 71-72, 81, 87, 89, 92, 282
Breckenridge, Iain 255
Brett, Arthur 111
Bruce, Lewis Knight 17, 29
Bute, Marquess of (and Family) xii, 1, 3, 8-9, 12, 20, 40-42, 44, 48-50, 64, 67, 76-77, 79, 199, 211, 215, 285
Byass, Sidney 105

Cartwright, William 27-29
Chalk, Charles 35-38, 45-47, 53
Clay Family (of St. Hilary) 61-62, 100-102, 118, 130, 168, 215-218, 240, 249, 282
Cooke, Nicholas 244-246, 257, 279
Corbett, Edwin 55
Cory Family 75-76, 81, 102, 109, 176, 215-216, 247, 249
Cosslett, John x, 256-257
Courtis, John 94, 120
Cousins, Harry 123-124, 156
Crawshay, William 22, 176, 222
Crosby, David 239, 242, 250
Cross, Edna 205-206, 213, 246

Danby, Morton 155
David, Charles 17-19, 26-27
David, Edmund 31, 50-51, 125

Davies, Alun 250, 269, 275
Davies, Kenneth 242
Deamer, Frederick 179
Dewey, Peter 279
Downing, Ivor 92, 123-124, 145, 149-151, 155
Duncan Family 111, 125-126, 136

Edwards, Bob x, 245, 279
Embiricos, Alex x, 269, 271, 279
England Family 121-122, 145
Ensor, Henry 94
Estevez, Camilo 207
Ewbank, Thomas 73-74

Feld, Val 241
Fisher, George 16, 19
Fletcher Griffiths, Bruce 231
Forrest, Robert 31
Forsdyke, Mair x, 207, 218, 255

Gagg, Roger 267
Gallagher, Paddy 255, 265
Gaskell Family x, 91-94, 145, 213
Gibbon, Samuel 17, 31, 87
Gill, Bill 249, 253
Glover, Ted 131, 179-180
Grubb, Lee 207

Hallinan, Lincoln 145, 235
Hancock Family 65, 82, 91-92, 180
Hann, Edmund 138
Harvey, William R. 46-47, 52-53
Heard, Henry 11-24, 28-32, 46
Heard, Percy 23-24
Heath, Thomas 15
Herbison, William 77, 104
Hill, Edward Stock 61, 71
Homfray Family (of Penlline) 6-7, 26, 30, 36, 87, 102, 110-111, 113, 120, 122, 149, 216
Howard, Sidney 15
Howell, Dan 207, 209
Howell, Edmund 73

Howell, James 35
Hughes, Robert 77
Hughes-Davies, Colin 217

Ingledew Family 78-82, 86, 109, 123-124, 282
Inkin, Geoffrey 237-240, 265
Insole Family (of Ely Court) 15, 20, 55, 63, 110, 114

Jackson, Howard 265
James, Keith 265
Jenner Family (of Wenvoe) 17, 73-74,
Jenour, Maynard 161, 186, 200-206, 211, 213, 215, 251
Johnson, Clive x, 285-286
Johnson, Sally 207
Jones, David Mansel ix, x, 214, 248, 251, 279
Jones, Helen Mary 236
Jones, John Elfed 236
Jones, John Price 50, 81
Jones, Mark 204, 207, 279
Jones, Michael 256-257, 267-269, 279-280
Jones, Robert Oliver 26
Jones-Evans, Dylan 274

Keen, Charles 145
Kinnock, Glenys 238-239
Knight, Edwin 19

Larby, Adrian 255, 271
Lawrance, Nick x, 281
Lazarus, John x, 151, 255, 263, 270, 271, 279
Lewis, Harry 271
Lewis, Henry and Family 31, 109-110, 137, 186, 223-225
Lewis, Ivor 127
Lewis, William H. 53-54
Lindsay Family (of Woodlands and Ystrad Fawr) 31, 69-70, 95-100
Lloyd-Edwards, Adrian 205
Lloyd-Edwards, Norman 211

Mackintosh, The Mackintosh of 30, 87-88,
MacWilkinson, Jeff 279
Mainstone, Keith 242, 249-250

Martyn, Dennis 176
Mathias Family 113-119, 146, 172-173, 252-253, 255
Matthews, Jack 218-222
Morgan, Eluned 236, 238
Morgan, John 35, 46, 52
Morgan Family (of Tredegar Park) 2, 6, 12, 20, 29-31, 36-37, 69, 96, 101, 135

Nash, Samuel 15
Newton, Jacob 53
Nicholl Family (of Merthyr Mawr) 90
Norris, Robbie x, 122, 142, 252

Ohlsen, Lars 15
Ollivant, Alfred 5
Orlebar, Evelyn 70, 95

Page, Charles 16
Paine, Henry 17
Parsons, Edmund 53
Payn, John 218, 255-256, 265
Peel, Fiona x, 151, 275, 277
Penn, George 155, 178
Phillips, Horace C. 148
Phillips, Peter 256
Price, Eddie 149
Price, Samuel J. 112
Preece, Ceri x, 75, 253, 267, 270, 279-280, 285
Prisk, Glenvin 280
Prosser, Ernest 106

Reardon-Smith, William 62
Reece, Edmund 45
Reece, Richard 17
Rees, Arthur 169
Rhys-Williams, Rhys 128, 148-149
Richards, Menna 238
Robert, Hywel ap 252
Robertson, Ally 194, 203, 213, 218
Rosser, Julian 274
Rosser, Melvyn 244
Rowe-Beddoe, David 236

Said, Dawn 207
Sainsbury, Jeff 212, 229-230, 237-238
Sansom, Arthur 195

Sennitt, Owen x, 278-279, 283
Seward, Edwin 30, 53
Shepherd, Leslie 231
Shepton, Samuel 56
Shewbrook, Henry 79
Shirley Family 69-70, 88, 95, 135
Simpson, Joe 126-127
Sloper, John 19
Spencer, Richard E. 15-16
Stacey, Cyril 5-6
Stacey, Frank 6-7, 40-42, 44, 49, 50
Staniforth, Joseph M. 11, 50, 93, 135
Stradling Griffiths, John 263
Stratton, Richard 31
Sweet-Escott Family 80-82, 86,

Tantum, Derek 203-204, 207, 218-219, 244
Taylor, Lionel E. 112
Thomas, George (of The Heath) 31, 97
Thomas, Gill x, 207
Thomas, Herbert 73
Thomas, H. Hugh x, 140
Thomas, John Griffin 109
Thomas, Lee 207
Thomas, Percy 156, 162, 178
Thomas, Roger 240
Thompson, Charles 135
Tipple, Edward 178-179
Traherne, Cennydd 87, 158-159, 161, 179, 214-215, 242
Turnbull, Maurice 130131, 143, 145, 151-153, 161, 164, 167-168
Turner, John 197, 204
Tyler Family (of St. Hilary) 31, 54-55,

Vaughan-Thomas, Wynford 175-177

Wain, Richard 18
Wakelin, Peter 242
Walters, Graham 280
Ware, James 20
Waring, Thomas 25-26, 32, 49
Warwick, Tim 257
Watson James, David x, 241-242, 250-251, 255, 263-264, 270, 279
Watkins, Tasker 164-165, 225, 229, 232, 239-240, 265
Watkins, William B. 18-20
Webb, Henry 110
Webb, John 197
Webber Family x, 119, 127-128, 131, 149, 156-157, 179, 187, 202, 225
Weston Evans, Arthur ix, 219, 256
Wilcox, Peter 207-208
Wilkins, Howard 267
Williams, Andrew x, 271, 279,
Williams, Angela 207, 257, 280
Williams, Bleddyn 261-263
Williams, Charles 16-17, 26-27, 29-31, 54
Williams, David 242
Williams, David Watkin 37-38
Williams Family (of Bonvilston) 17, 54, 59, 87, 102, 108, 109, 129, 130, 179, 193, 215
Williams, George C. 27, 53, 55
Williams, Griffith 256
Windsor-Clive Family (The Earl of Plymouth, of St. Fagans) 20, 31, 70, 86, 167-168
Wooller Family x, 148, 166-170, 172, 183, 221, 224-225, 252, 267, 284
Worthington, George 6-7, 17-17, 27-28, 45
Wynne-Jones, G.V. 119, 172-174, 179, 205, 213, 262

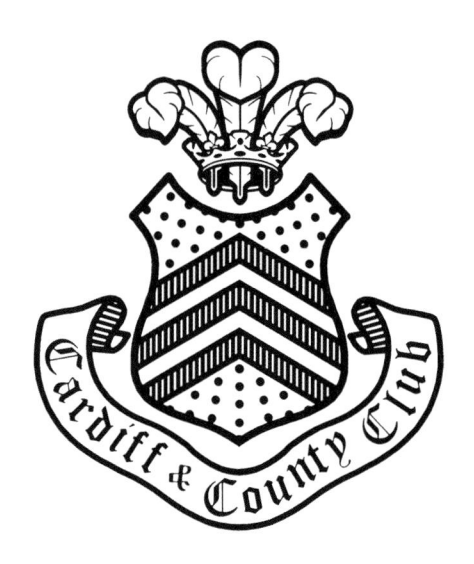

Members' Supplement

This special supplement to *Always Amongst Friends* contains additional features and illustrations unique to the edition published only for members of the Cardiff and County Club and has been prepared with the assistance of Clive Johnson, Chairman of the Club's Anniversary Sub-Committee, and Ceri Preece, the Club's Honorary Secretary.

1. The Cardiff and County Club's 150th Anniversary Celebrations
2. Founder Members of the Cardiff and County Club – 1866
3. Members of the Cardiff and County Club – during the 150th Anniversary Year
4. Presidents of the Cardiff and County Club – 1866-2016
5. Senior Officers and Committee Members – during the 150th Anniversary Year
6. The Cardiff and County Club's Art Collection

1. The Cardiff and County Club's 150th Anniversary Celebrations

Always Amongst Friends is a most apt title not just for this book and the Club as a whole, but also for the planning and organisation surrounding the 150th anniversary celebrations. A small committee had been created in September 2014 to organise the anniversary year, with myself as Chairman, as well as Club Chairman Owen Sennitt, Alex Embiricos, the immediate Past Chairman, Ceri Preece the Club's Honorary Secretary, Christopher Dale, Roger Gagg, the late John Adams and Nick Lawrance, the Club's General Manager.

It was agreed at an early stage that we wanted a fun and memorable year with a number of highlights and at least three major events.

- The first – to mark the signing of the agreement to found the Club in January 1866;
- The second – to mark the opening of the first Clubhouse in St. Mary Street, adjacent to the Royal Hotel on April 2nd, 1866; and
- The third – to celebrate in grand style with a Gala Dinner and to raise money for a charitable cause.

It was also agreed that a letter be written by Ceri Preece, our Honorary Secretary, to our Patron, His Royal Highness The Duke of Edinburgh, inviting him to visit the Clubhouse during our anniversary year. We were delighted to receive a response, in September 2015, indicating that our Patron would consider a visit once he and his Private Secretary had confirmed an itinerary for the year.

Much of our early deliberations for the Gala Dinner centred on the offer from the internationally acclaimed and award winning soprano Rebecca Evans, wife of one of our trustees, Stephen Jones, to perform with her neighbour and good friend Sir Bryn Terfel, the renowned bass baritone and a Club member to boot! With the capacity of the Clubhouse's dining room at around 70, we soon realised that we needed a venue with a larger capacity. Initial thoughts and a great deal of work were expended in investigating and costing out the installation of a marquee on the car park. But at around £15,000, this proved not only far too expensive, but ran the risk of clashing with an event in the newly renamed Principality

Stadium. In the event, there was a huge concert by Welsh rock band, The Stereophonics, in town on the same evening, Saturday June 4th, but at the football stadium in Leckwith and not next door.

Everyone wanted a memorable evening so a range of alternative venues were investigated, including the Royal Welsh College of Music and Drama, the stage of the Wales Millennium Centre, the City Hall and the National Museum of Wales. After careful consideration the option of holding a function at the National Museum was chosen, especially as it had sufficient capacity and the franchised catering company, Elior, was very happy to work with the Club's staff and a guest chef. Soon afterwards, we were delighted to find that the renowned chef, Stephen Terry, of The Hardwick near Abergavenny was willing to come and lead the team at our 150th celebrations. Stephen had been one of our Guest Chefs during 2015, and had experienced working with the Club's kitchen team to produce a meal of the highest quality.

In the meantime, Tim Mathias – a long-standing member of the Club and renowned tie designer – was invited to design a tie to mark the anniversary. He excelled and, in typical fashion, recommended not one but two ties, one bold in Club blue with gold stripes, the other more sombre with just the Club's crest and the letters CL, a scholarly indication of the cause for celebration. The first outward sign that things were progressing came when the ties arrived and members were given the choice of which one to have.

Hot on the heels of the discussions over the venue for the Gala Dinner and the design of the tie was the decision to have a special anniversary wine. Members of the 150th committee, plus the wine and social committees, put a great deal of effort into selecting an appropriate claret – in fact they were so diligent that several members complained of repetitive palate strain! An order was duly placed for delivery in the autumn of

D. Hugh Thomas, Alun Thomas, D. Ken Jones and Tony Ball attending the St. David's Day Dinner in 2016.

2015, and a label designed with the help of printer and former committee member, Stuart Brookes, of McLays. It had been agreed, in 2014, to invite Dr. Andrew Hignell to write this book, reflecting on the Club's development over the course of the last 150 years and a publication where there would be a permanent record of the Club's celebrations during the anniversary year. In the event, Andrew has not only produced this excellent history of the Club, but has cleverly woven it into the history of Cardiff, and in particular, its changing economy.

Such a programme of events and celebration is rarely without problems and delays, and having chosen an anniversary claret months earlier, at the last minute our supplier told us that he had labelled our wine for another customer and offered an alternative. Sadly, this didn't come up to expectations, so we turned adversity into opportunity by choosing seven anniversary wines, enabling us to choose wines to suit a variety of tastes.

The Allan Scott Cecilia Brut from New Zealand was our sparkling choice, supported by three red and three white wines. Our friends and suppliers, particularly Nigel O'Sullivan from Fine Wines Direct, and Amanda Kynaston of The Bay Tree Wine Company were particularly helpful. One red came from the Southern Rhone, in the form of a lirac from Moulin des Chenes, the second was an excellent claret, from Sean and Nicola Allison, whose parents Bob and Sue Watts, had established the Chateau de Seuil vineyard on the banks of the Garonne in 1988. For a lighter wine, a Pinot Noir was chosen from Allan Scott in New Zealand, along with his Sauvignon Blanc in the white range, A Viogner from the Languedoc and a Spanish Albarino completed our selection.

After 15 months in the planning, our anniversary year got off to a fine start on January 3rd, 2016, with over 100 members and guests attending a lunch in the Clubhouse to mark the date of the agreement to found our Club. Allan Scott Cecilia Brut was served to the sound of a brass quintet on the grand staircase landing. Head Chef Lee Thomas prepared

Christopher Dale (left) and Robbie Norris (right) enjoying the Champagne Reception, held in the French Impressionists gallery, prior to the 2016 Gala Dinner at the National Museum of Wales.

the superb menu, and the assembled company joined in a toast to 'Our Founders' proposed by Chairman, Owen Sennitt. Other events quickly followed during 2016, notably the *Burns Nicht* on January 28th, where Professor John Lazarus proposed a toast to the immortal memory of the Scottish poet whilst, on March 1st, a dinner was held to celebrate Wales' Patron Saint, St. David.

During the late autumn of 2015 we received confirmation that H.R.H. The Duke of Edinburgh was intending to join us for lunch on Wednesday March 16th, 2016 and, after conversations between Ceri Preece and H.R.H.'s Private Secretary, a set of guidelines for managing the Duke's visit were received and, after members had been informed that our long-standing Patron would be visiting the Clubhouse in Westgate Street, there was a *frisson* of excitement with some rapid and detailed planning.

We knew that demand for places at this function with Prince Philip would be high, but we soon realised that there would be a limit to the number of diners

The Club's Patron, H.R.H. The Duke of Edinburgh, arrives at the Clubhouse

Vice President, Paul Twamley (left) and President, David Mansel Jones (centre) extend the warmest of welcomes to H.R.H.

Sir Norman Lloyd-Edwards (left) shares a word with the Patron and the President.

we could accommodate in the main dining room plus the Westgate Room. With around 100 places on offer, it was decided to hold a ballot to determine not only who should attend the Patron's Lunch, but also in which room they would be seated. We were very pleased as well that His Royal Highness was keen to meet all of the members attending the function, plus the accompanying staff. So, after several discussions, this was achieved by some expert choreography and arranging a series of small groups around the Clubhouse, including the committee and reading rooms on the second floor, where our Patron could meet and mingle with members.

The excellent menu for the lunch was conceived, designed and delivered by Head Chef, Lee Thomas, with anniversary wines to suit. His Royal Highness then proposed a toast to the Cardiff and County Club, accompanied by a glass of Glamorganshire Pale Ale. He was also very generous in his comments, and donated a new Visitors Book, which he presented and signed during the visit. Particular thanks go to our Honorary Secretary, Ceri Preece, who spent many hours to ensure that the Duke's visit was a success.

Close on the heels of the visit by our Patron was an Anniversary Dinner for members to mark the opening of the first Clubhouse on April 2nd, 1866. At what was a members-only event, 90 diners were present as our President, David Mansel Jones, proposed the toast to the Club.

A further highlight was the Gala Dinner on June 4th. The build up to this event was particularly exciting and, with a final tally of 264 guests, a fantastic evening was held in the Grand Hall of the National Museum of Wales, with a sparkling wine reception in the fabulous impressionist galleries, supported by the delightful playing of the harp by Hannah Stone.

The diners then moved into the Grand Hall, where Rebecca Evans started the entertainment, accompanied by leading Welsh accompanist Jeffrey Howard. Bryn Terfel then gave a fabulous rendition of the lullaby *Suo Gan*, followed by *If I Were a Rich Man* which rivalled that of Chaim Topol in the film *Fiddler on the Roof*. After several duets, including a medley from George Gershwin's *Porgy and Bess*, there was spontaneous and most well-deserved applause where our guests stood to congratulate the duo on a truly fabulous performance. The artistes also returned to the stage as Lisa Williams, who had been overseeing the administration of the event, presented a bouquet to Rebecca, and bottles of Welsh Whisky to Bryn and Jeffrey.

The evening also saw Ceri Preece, as the Club's Honorary Secretary, say grace before everyone enjoyed a fabulous meal prepared by a team led by Stephen Terry, working closely with Lee, Dan and Cam from the Club, before Lord Lieutenant Dr. Peter Beck gave the Loyal Toast. The Gala Dinner had also been designed to raise money for the charity Tŷ Hapus, nominated by both Bryn and Rebecca. This charity supports people and families afflicted

H.R.H. The Duke of Edinburgh with Officers of the Club and the Lord Lieutenant during the Patron's visit to the Club in March 2016. Left to right: Andrew Williams, Paul Twamley, Stephen Jones, Clive Johnson, David Mansel Jones, the Patron, Dr. Peter Beck, Owen Sennitt, Lawrie Williams and Ceri Preece.

His Royal Highness completes his witty toast to the Club at the conclusion of the Patron's Lunch.

Members David Lermon, David Williams, John Roberts and John Phillips at the Anniversary Dinner.

by dementia, and President David Mansel Jones was very pleased to present a cheque on the night to Rebecca and Stephen, both trustees of the charity, for £6,000.

Particular thanks are due to the following friends of the Club who donated the raffle prizes, including Shaun Hill of The Walnut Tree, James Sommerin of the eponymous restaurant, Peter Dewey of Town and Country Hotels, Margaret Waters of the Park Plaza Hotel, Anand George of The Purple Poppadum, Christopher Dale of the Firing Line Museum, Dudley Newbery of The Miskin Arms, and our beer supplier, InBev.

2016 was, therefore, a memorable year for all those involved in the Club's celebrations. We have left a fitting legacy for the Club, and hope that those involved in the 175th anniversary celebrations (as well as those involved in the bicentenary) have as much pleasure as we have had in arranging such a fantastic year of social events. As this splendid book recounts, the Club and its members – both young and old – can look forward to a very rosy future as the 21st century further unfolds.

Clive Johnson　　　　　　　　　　　　　　　　　　　　　　**Ceri Preece**
Chairman, 150th Anniversary Committee　　　　**Honorary Secretary**

Club members and honoured guests enjoying the 2016 Gala Dinner. Back row (left to right) Sir Bryn Terfel, Ross Griffin, Rebecca Evans, Malcolm Puntis and Mrs. Madley. Front row (left to right) Hannah Stone, Stephen Jones, Elena Griffin, Paul Madley and Mrs. Puntis.

2. Founder Members of the Cardiff and County Club – 1866

Adams, William (Roath)
Alexander, William (Cardiff)
Allen, William (Llwynarthen)
Allen, Richard (Roath)

Ballard, Captain (Cowbridge)
Bassett, Alexander (Llandaff)
Bassett, Richard (Bonvilston)
Batchelor, John (Cardiff)
Batchelor, Sydney (Penarth)
Bath, Charles (Swansea)
Bernard, Charles (Cardiff)
Bird, Henry (Cardiff)
Bird, John (Cardiff)
Blosse, The Venerable Archdeacon (Bridgend)
Booker, John (Cardiff)
Booker, Thomas (Cardiff)
Boyle, John (London)
Bruce, Lewis (St. Nicholas)
Bute, The Most Noble Marquis of (Cardiff)

Carne, John Nicholl (St. Donats)
Clark, George (Dowlais)
Coleridge, Charles (London and Haverfordwest)
Corbett, John (Cardiff)

David, Charles (Cardiff)
David, Edmund (Radyr)
David, Edward (Fairwater)
Davis, Lewis (Cardiff)
Dawson, James (Bridgend)
Dobson, Samuel (Cardiff)

Ensor, Tom (Cardiff)
Evans, Henry (Cardiff)

Fisher, George (Cardiff)

Gibbon, Samuel (Cowbridge)
Giffard, Henry (London)
Goddard, Thomas (St. Fagans)
Goddard, William (Cowbridge)
Gould, Captain (Cowbridge)
Grainger, Frederick (Cardiff)
Greenhill, Frederick (Roath)
Griffith, Robert (Cardiff)
Grover, Montagu (Cardiff)

Harvey, William (Cardiff)
Heard, Henry (Cardiff)
Heath, Thomas (Cardiff)
Hemingway, James (Cardiff)
Hemingway, John (Cardiff)
Herbert, James (Ross-on-Wye)
Hewitt, John (Tir Mab Ellis)
Hill, Edward (Llandaff)
Hill, George (Cardiff)
Hodge, Thomas (Cardiff)
Homfray, John (Penlline Castle)
Huckwell, Joseph (Lavernock)

Insole, James (Llandaff)
Insole, Walter (Llandaff)

Jenkins, William (Caerleon)
Jenner, Robert (Wenvoe Castle)
Jones, Captain (Fonmon Castle)
Jones, Robert (Fonmon Castle)

Kemys-Tyne, Colonel (Cefn Mably)
Knight, Rev. Charles (St. Brides)

Lee, Major (Lanelay Hall)
Lewis, Wyndham (The Heath)

Lewis, Rev. Charles (Peterston-super-Ely)
Lewis, Evan (Aberdare)
Lewis, James (Aberdare)
Lisle, Robert (Cardiff)
Llewellyn, William (Bridgend)
Luard, Charles (Llandaff)
Lucas, Clement (Cardiff)

Michael, William (London and Cardiff)
Miller, Thomas (Llandaff)
Morris, James (Cardiff)
Morgan, The Hon. Frederick (Ruperra Castle)

Nash, Samuel (Cardiff)
Nicholl, George (Cowbridge)
Nicholl, John (Merthyr Mawr House)

Ohlsen, Lars (Llandaff)

Page, Charles (Llandaff)
Paine, Henry (Cardiff)
Palmer, Captain (Llandaff)
Pearson, Charles (Roath)
Powell, Thomas (Newport)
Prichard, William (Cowbridge)
Pride, James (Cardiff)

Raby, William (Cardiff)
Rawlinson, George (Bath)
Rea, Robert (Pontypridd)
Reece, Edmund (Cardiff)
Reece, James (Cardiff)
Reece, Robert (Cardiff)
Richards, Major (Cardiff)
Rickards, Rev. Hely (Llandough)

Salmon, Thomas (London and Cardiff)
Saunders, Joseph (Cardiff)
Sheen, Alfred (Cardiff)
Shirley, Lewis (Cardiff)
Sloper, John (Cardiff)
Spencer, Richard (Llandaff)

Stacey, Rev. Cyril (Whitchurch)
Stephens, Thomas (Cardiff)
Strick, James (Swansea)
Stuart, Charles (Cardiff)
Stuart, Major General (Cardiff)

Taylor, William (Cardiff)
Thomas, Andrew (Roath)
Thomas, George (Ystrad Mynach)
Thomas, Herbert (Llanblethian)
Thompson, Astley (Whitchurch)
Thompson, Charles (Cardiff)
Thompson, George (Cardiff)
Traherne, Arthur (Bridgend)
Tredegar, Lord (Newport)
Tyler, Colonel (Cotterill Park)

Vaughan, Edward (Neath)
Vivian, Hussey (Swansea)

Waldron, Clement (Llandaff)
Wallace, Samuel (Cardiff)
Waring, Thomas (Cardiff)
Ware, James (Cardiff)
Watkins, William (Cardiff)
Watkins, Thomas (Cardiff)
Watson-Jones, Colonel (Cardiff)
Williams, Charles (Roath)
Williams, Rev. David (Pontypridd)
Williams, Edwin (Duffryn Ffrwd)
Williams, Evan (Duffryn Ffrwd)
Williams, G Crofts (Llanrumney)
Williams, George (Hendredenny)
Williams, Gwilym (Miskin Manor)
Williams, Edgar (Pwllypant)
Williams, Richard (Cardiff)
Williams, Thomas (Bonvilston)
Williams, W. Lloyd (London and Cardiff)
Williams, William (Pontypridd)
Woods, Rev. George (Sully)
Woods, Thomas (Cardiff)
Worthington, George (Cardiff)

3. Members of the Cardiff and County Club – during the 150th Anniversary Year

Life Members

The Earl of Plymouth – Windsor-Clive, O.R.I.
Payn, Commander J.E. RD RNR

Abbott, S.R.
Adams, J.J. MBE*
Aggarwal, R. OBE, DL
Alfaham, Dr. M.
Allen, N.J.
Anderson, M.H.C.
Andrews, Prof. J.A. CBE
Anthony, G.
Antoniazzi, J.L.
Arter, J.J.C.
Arter, J.J.M.
Ayling, G.T.

Baggott, M.J.
Ball, A.K.
Barnett, E.R.
Barnwell, M.A.T.
Barry, J.D.
Bartlett, J.G.
Baxter-Wright, T.R.
Baynham, R.W.
Beard, Col. N.R. DL
Beck, Dr. P. CVO, MD, FRCP
Bell, P.D.
Bennett, T.H.
Berridge, A.G.
Berry, S.R.
Bevan, D.M.
Bevan, D.M.
Biddle, Dr. N.M.
Biles, A.J.
Biles, R.C.
Bishop, M.L. QC

Board, A.G.
Bohana, R. MBE
Bond, D.
Bowen-Williams, N.
Bowles, P.M.
Braddick, R.*
Brain, C.M.
Brain, C.N.
Breckenridge, I.M.*
Brice, I.C.
Brookes, S.J.C.
Brooksbank, Prof. D.J.
Bryon, K.J.
Bugler, M.J.
Bulmer, H.W.
Burch, J.I.
Burch, R.J.

Cadogan, Sir J.I.G. CBE Kt
Calcaterra, R.J.
Cammish, R.A.
Cantlay, Major D.A.
Capel, M.
Carey-Evans, W.L.
Castledine, S.I.
Catris, M.B.
Chapman, Prof. A.J.
Chapman, J.R.
Charles, Major J.E.
Chaston, R.A.
Child, E.M.
Child, H.
Chronik, N.

Clark, Dr. R.
Clarke, C.L.D.
Clarke, Lt. Col. D.G. OBE
Clay, M.J.
Clay, R.M.
Cleaver, G.J.
Clements, A.S.
Clyne, G.R.
Cockbill, Dr. S.M.E.
Cody, B.A.
Cole, R.D.
Coleman, P.F.
Coliandris, T.
Colston, A.I.
Colston, I.E. JP
Conway, D.
Coombes, J.
Corrigan, P.A.*
Cosslett, J.S.
Cozens, G.J.
Crawshay, H.H.R.
Cresswell, C.T.
Crewe-Read, Lieut. Col. N. MBE
Crookes, R.F.
Crosby, W.M.L.
Crowley, J.M.
Curran, His Honour J.T.
Curtis, S.L.
Cutlan, J.

Dale, C.F.
Daniels, A.M.
Daniels, P.C.T.
Daniels, T.J.P.
Davenport, J.R.P.
Davey, J.L.
David, Dr. B.
David, C.M.
Davies, A.C.G.
Davies, A.J.
Davies, B. OBE
Davies, C.J.G.
Davies, D.
Davies, E.M.
Davies, Dr. H.F.

Davies, H.G.
Davies, H.W.
Davies, J.A.
Davies, J.D.R.
Davies, J.G. CBE
Davies, J.G.
Davies, J.H.
Davies, Wing Comm. J.I. CBE
Davies, J.J.T.
Davies, J.L.
Davies, J.L.
Davies, J.P.
Davies, Prof. Col. K. MBE, RRC, TD, DL
Davies, L.J.
Davies, L.S. JP
Davies, M.J.
Davies, M.J.
Davies, P.E.C.
Davies, Dr. P.M.
Davies, P.T.
Davies, R.G.
Davies, R.H.
Davies, R.J.
Davies, S.J.T.
Davies, W.H.W.
Davies, W.P.
Dawson, M.V.
De Lloyd, D.N.
De Lloyd, W.R.
Dellar, J.W.
Dellar, P.G.
Dewey, P.R.
Dobson, A.J.
Dobson, P.
Dodd, J.H.
Dolan, P.D.B.
Dollery, P.N.
Donovan, D.M.
Downing, J.C.R.
Doyle, B.E.
Dykes, Dr. D.W. K St. J

Eddershaw, M.C.
Eddins, P.F.
Edmunds, J.A.M.

Edwards, H.M.
Edwards, J.T.
Edwards, R.D.
Edwards, R.N.
Elias, C.S.
Elias, G. QC
Embiricos, A.P.
Evans, A.G.
Evans, Dr. C.E.
Evans, D.A. QC
Evans, G.E.
Evans, G.W.W.
Evans, G.R.W. QC
Evans, H.R.E.
Evans, J.M.
Evans, J.P.
Evans, J.W. CBE
Evans, L.J.
Evans, M.A.
Evans, N.P.W.
Evans, R.A.
Evans-Bevan, T.R.
Eveleigh, G.W.

Fairweather, C.P.
Fairweather, I.M.
Fitzherbert, A.T.
Fitzwilliam, D.J.L.
Fletcher, A.P.
Fletcher, S.B.
Forbes Keir, J.A.
Ford, D.J.
Ford, E.I.
Ford, R.L.
Forster, R.A.
Fox, A.P.C.
Furness, H.J.
Furness, His Honour M.R.

Gagg, R.D.
Gamble, Brigadier M.J.
Gaskell, His Honour J.W.
Gibbon, D.F.*
Gibbons, J.F.
Gibbs, N.C.

Gibbs, N.G.
Gibbs, P.D.E.
Gilbart, J.L.
Gilbertson, C.E.M.
Giles, A.D.
Giles, H.R.C.
Gill, W.A.
Glaister, K.D.
Golley, A.O.
Gooding, R.A.F.
Graham, E.P.
Gretton, A.L.
Grey, S.C.P.
Griffin, R.T.
Griffith, N.J.C.
Griffith Williams, The Hon. Sir J. QC
Griffiths, A.W.
Griffiths, G.
Griffiths, J.
Griffiths, R.M.
Griffiths, J.P.G. QC
Groves, D.D.
Gunther-Bushell, Dr. M.

Haines, K.V.
Hall, I.C.
Ham, J.
Ham, R.W.
Hancock, G.T.
Handyside, R.G.
Hansen, J.D.
Harding, P.W.
Harmond, I.G.
Harrington, Dr. M.J.
Harrington, P.J. QC
Harrison, R.J.M.
Harry, Lieut. Col. K.R.
Harvey, Revd. Dr. D.V.R.
Harvey, W.H.
Hawkins, D.G.
Hayes, F.E.S.
Hayes, J.F.R.
Hayward, A.
Heale, A.G.
Henderson, Prof. A.H. OBE

Hennessy, Dr. N.J.
Hensman, J.H.G.
Herbert, D.R.J.
Hill, D.
Hill-John, G.M.
Hodge, K. OBE
Hodgson, Dr. A.
Holcombe, Canon G.W.A.
Holmes, J.F.
Honeyman, D.A.
Hood, M.L.
Hooker, M.J.
Hooper, R.J.
Hooper, S.C.
Hopkins, C.E.
Hopkins, J.H.D.
Horley, T.G.
Howard-Cook, S.J.
Howell, A.R.
Howell-Pryce, B.
Howells, G.W. MBE
Howorth, T.R.
Hufton, N.R.
Hughes, G.O.
Hughes, M.A.
Hughes, P.E.
Hughes, R.J.
Hughes-Davies, P.A.
Hughes-Lewis, D.W.
Hughes-Lewis, J.
Huish, J.V.
Humphries, N.J.
Hunt, W.
Hurrell, N.R.
Hutchings, R.J.
Hutton, A.D.

Ingledew, J.F.
Ingleson, M.P.

Jackson, M.R.
Jacob, I.
James, D.K.M. OBE
James, H.J. Order St. James MBE
James, J.M.

James, M.E.
James, M.R.S.
James, R.A.
Jardine, P.J.
Jarvie, T.H.
Jasinski, R.
Jenkins, D.W.
Jenkins, E.R.
Jenkins, R.J.
Jenkins, S.C.
Jeremy, A.W.
John, P.D.H.
John, R.L.
Johns, S.R.
Johnson, D.C.
Johnson, Col. N.A. OBE TD FIMI DL
Jonathan, D.M.
Jones, A.P.
Jones, A.W.
Jones, B.D.
Jones, B.K.
Jones, B.T. CBE
Jones, B.W.R.
Jones, D.C.
Jones, D.G.
Jones, D.K.
Jones, D.M.
Jones, D.R.
Jones, E.H.
Jones, F.C.
Jones, G. OBE
Jones, G.H.
Jones, His Honour G.J.
Jones, G.J.
Jones, H.C.
Jones, His Honour H.D.H.
Jones, H.G. CBE
Jones, H.W.
Jones, J.D.
Jones, Dr. J.E. CBE
Jones, J.M.
Jones, M.A.
Jones, M.H.
Jones, M.I.P.
Jones, N.M.

Jones, N.P.
Jones, P.H.M.
Jones, Dr. P.W.
Jones, R.A.
Jones, R.L.
Jones, Sir R.S. Kt. OBE
Jones, R.W.A.
Jones, S.D.F.
Jones, S.R.
Jones-Evans, Prof. D. OBE
Jordan, M.G.

Kelsall, I.M. OBE*
Kember, R.
Kendall, R.S.
Kenney, G.D.
Kiddy, A.
Kilmister, Lieut. Col. J.C.
Knight, Dr. A.G.
Knight, R.H. JP
Knowles, T.
Kynaston, Prof. H.G.

Larby, A.G.B.
Larby, C.A.
Lawley, M.A.
Lazarus, Prof. J.H.
Leach, G.O.
Leach, H.M.
Lermon, D.N.
Lewis, A.L.P.
Lewis, D.W.
Lewis, H.G.
Lewis, O.P.
Lewis, R.C.
Lewis, R.H. MBE
Lewis, R.J.
Lewis, R.J.A.
Lewis, T.K.
Lindsay, R.E.
Lister-Sims, S.
Little, D.E.
Llewellyn, M.E.R.
Llewellyn, R.J.
Llewellyn-Jones, A.M.

Lloyd, G.C.
Lloyd, Dr. H.J.
Lloyd, N.C.
Lloyd Jones, Sir R.A. KCB
Lloyd-Edwards, D.R.A JP
Lloyd-Edwards, Sir N. KCVO Gc St. J Rd
Lock-Necrews, J.
Long, G. MBE
Long, M.V.
Loosemore, M.E.
Lynn, G.E.

MacWilkinson, J.N.
Madley, P.W.S.
Madley, R.J.B.
Maheson, M.
Manners, A.J.
Mansel Lewis, R.
Manuel, R.W.
Marsh, B.
Marshall, A.R.
Martin, Dr. R.
Mathias, T.F.
May-Hill, R.C.
May-Hill, R.R.A.
Maynard, M.P.
Mayne, W.D.
McCarthy, D.
McGrane, M.E.
McLean, J.M.
Meek, A.M.
Meirion-Williams, R.J.
Meredith, A.D.
Metcalfe, L.C.
Milford, A.I.
Millward, M.E.
Mitchell, J.D. OBE
Mitchell, J.P.
Moore, J. MBE
Moreton, A.J.
Morgan, Prof. B.
Morgan, Judge D.H.
Morgan, D.R.
Morgan, D.W.C. OBE

Morgan, J.P.W.
Morgan, J.K.
Morgan, Dr. J.R.
Morgan, M.D.
Morgan, M.J.G.
Morgan, N.L.T.
Morgan, R.J.
Morgan, R.M.
Morgan, R.P.
Morgan, R.W.
Morgans, G.W.
Morgans, J.M.
Morris, A.W.
Morris, His Honour D.G.
Morris, G.J.C.
Morris, S.B.
Morris, W.D.
Mossford, A.C.E.
Mullen, D.B.
Murphy, D.M.
Murphy, I.P. QC
Myrddin-Evans, D.G.

Needham, P.J.E.
Newbery, D.
Newton, I.S.
Ng, S.A.
Ng, Dr. W.S.
Nicholas, M.G.
Nicholls, J.L.
Norman, L.T.
Norris, Major C.R.V.

Osborne, D.C.
Osborne, G.C.
O'Sullivan, D.B.P.
O'Sullivan, N.
Owen, A.E. OBE
Owen, I.H.M.
Owen, J.C.D.
Owen, Rev. J.T.A.
Owen, J.W. CBE
Owen, R.E.

Paddison, D.F.

Palmer, M.S.
Palmer, R.W. CBE
Parker, A.G.
Parkinson, R.T.
Parry, T.M.
Parsley, C.R.
Parsons, A.D.
Parsons, R.
Partridge, G.
Patterson, Col. L.
Payne, Dr. I.
Peacock, R.F.
Pearson, A.J.K.
Peel, F.N.
Petersen, Major T.J. TD JP
Peterson, A.E.
Peterson, H.R.E.
Peterson, J.C.
Pexton, R.D.
Phillips, B.G.L.
Phillips, D.E.
Phillips, D.J.
Phillips, J.W.
Phillips, N.W.
Phillips, Sir P.J. OBE
Phipps, M.C.
Phipps, M.L.
Pickering, D.F.
Pill, H.R.
Pill, Rt. Hon. Sir M.T.
Porter, M.D.
Powell, K.J.
Powell, R.C.P.
Powell, R.H.
Powell, T.M.
Powis, R.J.
Preece, D.C.W.
Preece, D.J.A.
Preece, D.M.W.
Preece, M.T.D.
Prendergast, M.T.
Price, E.G.
Price, I.D.
Price, N.M.A.L.
Price, His Honour P. QC

Price-Davies, G.
Prichard, M.C.T.
Prior, M.J.
Pritchard, G.S.
Prosser, His Honour E.J. QC
Pugh, A.I.
Pugh, M.R.
Pugh, N.
Pugh, R.G.
Pugsley, R.J.C.
Puntis, Dr. M.C.A.

Ralph, Prof. B.
Ramadan, N.
Rawlins, J.C.
Read, R.M.H.
Reardon-Smith, J.P.
Reardon-Smith, R.W.A.
Reardon-Smith, S.J.
Reed, G.B.
Reed, J.W.
Rees, C.A.
Rees, H.J.
Rees, Dr. J.A.E.
Rees, J.D.
Rees, J.S.
Rees, P.R.
Rees, T.J.R.
Rees, T.P.H.
Reid, S.P.
Reid-Jones, A.
Renwick, A.N.
Richards, C.J.
Richards, H.L.P.
Richards, J.C.
Richards, His Honour P.B.
Richards, P.E.
Richards, R.O.A.
Roberts, D.G.
Roberts, D.T.
Roberts, J.A.
Roblin, C.E.
Roblin, J.
Roddick, G.W. QC
Rodger, P.R.C.

Rogers, P.D.C. TD
Rosser, J.

Sadka, B.E.J.
Sadka, S.I. OBE
Salway, D.
Sanders, J.
Sater, R.M.
Saunders, C.J.
Saunders, N.
Scale, M.J.
Scott, D.D.
Scott, P.D.
Scott, W.A.
Sennitt, O.S.A.
Sennitt, R.H.
Servini, R.
Seys-Llewellyn, His Honour A.J. QC
Shaw, M.P.
Shepherd, R.G.
Sievewright, K.C.
Sims, N.W. MBE
Skelding, S.
Slade, M.E.
Slatter, Sqd. Ldr. C.
Smart, J.R.
Smart, R.A.
Smith, R.H.
Spencer, B.J.
Stealey, D.
Stephanakis, C.C.
Stokes, D.J.
Strevens, C.J.
Strong, W.E.
Styles, D.W.
Sullivan, M.P.N.
Summers, R.A.
Suthers, Lieut. Col. J.D.

Taylor, N.B.
Taylor, P.W.D.
Taylor, T.T.W.
Temple-Morris, Lord P.
Theakston, T.J.O.
Thomas, A.

Thomas, C.R.
Thomas, D.E.W. OBE
Thomas, D.G.
Thomas, D.H. CBE
Thomas, D.P.G.
Thomas, D.T.
Thomas, G.W.
Thomas, H.H.
Thomas, H.O.
Thomas, Lieut. Col. H.W.I.
Thomas, J.F.
Thomas, K.C.H.
Thomas, L.F.G.
Thomas, M.
Thomas, P.D.
Thomas, R.H.
Thomas, R.J.
Thomas, Dr. S.J.
Thomas, Col. S.O. CBE
Thomas, R.G. OBE
Thompson, C.
Thorpe, B.
Thorpe, S.P.
Tillyard, J.H.H. QC
Tovey, G.S.
Toye, N.M.
Trigg, C.M.
Truman, K.D.W.
Tuck, J.F.
Tuck, M.
Tudor-John, W.
Tutssel, W.G.
Twamley, P.J.
Tyler, B.A.

Van Rees, Col. T.J. MBE ED DL
Vaughan, The Hon. J.E.M.
Vaughan, K.J.
Vaughan, R.F.
Vere-Whiting, C.G.
Verrier-Jones, Dr. E.R.
Vicari, Maitre A.*

Walker, A.J.
Walker, P.C.

Wall, G.G. JP
Walters, A.
Walters, D.M.
Walters, G.A.
Walters, R.J.
Ward, O.W.N.
Wardell, A.G.
Watkin, J.J.
Watkins, D.J.M.
Watkins, D.L.
Watkins, E.T. MBE JP
Watkins, J.K.G.
Watkins, O.
Watkins, O.R.
Watkins, R.D.H.
Watson, Dr. M.W.
Watson, R.P.F.
Watson James, D. OBE
Watts, S.J.
Weatherill, E.L.P.
Webb, J.D.
Webber, D.M. TD
Weber, J.S.
Westlake, N.D.
Wheatley, O.S. CBE
Wheeler, D.K.A.
White, C.S.
Whiting, R.
Wigley, I.T.
Wilkins, D.H.
Wilks, Major C.W.
Williams, A.B.
Williams, Dr. A.C.
Williams, A.C.
Williams, A.G.C.
Williams, A.J.
Williams, A.V.
Williams, B.M.
Williams, C.J.C.
Williams, C.P.R.
Williams, C.R.
Williams, C.S.N.
Williams, D.B.
Williams, D.L.
Williams, D.L.

Williams, D.N.O. OBE
Williams, G.E.
Williams, G.J.
Williams, G.M.
Williams, H.G.
Williams, H.R.C.
Williams, Dr. K.C.
Williams, L.
Williams, L.H.W.
Williams, Dr. M.A.H.
Williams, M.D.
Williams, M.O.
Williams, P.H.
Williams, P.J.
Williams, Sir P.M. OBE
Williams, R.A.
Williams, R.B.
Williams, R.D.
Williams, R.J. QC
Williams, R.J.E.
Williams, R.W.M.
Williams, T.E.
Williams, Sir W.L.
Windsor-Lewis, G.
Woodhouse, A.J.

Woodley, R.
Wooller, J.H.
Worrall, J.R.

Yapp, J.L.
Yapp, P.J.T.
Yeandle, A.C.
Yeandle, A.D.
Yendoll, H.D.
Young, B.C.
Young, M.G.
Young, Dr. M.H.

Zinni, V.

Honorary Members

Capon, The Very Reverend G.H.
Evans, Sir M. FRS
Gethin, D. OBE
Hignell, Dr. A.K.
Morgan, Rt. Revd. Dr. B.
Miller, Commodore A.J.G. CBE RN

*Deceased members – January 1st to December 31st, 2016.

4. Presidents of the Cardiff and County Club – 1866-2016

Richard Wyndham Williams	1866-1902
Charles Williams	1902-1908
Fred Insole	1908-1917
Henry Lewis	1917-1925
Herbert Richards Homfray	1925-1940
Rhys Rhys-Williams	1940-1950
Robert Webber	1950-1962
'Tip' Williams	1962-1974
Maynard Jenour	1974-1992
John Cory	1992-2003
Glynne Clay	2003-2011
David Mansel Jones	2011 to date

5. Senior Officers and Committee Members – during the 150th Anniversary Year

President	D.M. Jones
Vice-President	P.J. Twamley
Trustees	D.M. Jones
	L.H.W. Williams
	P.J. Twamley
	S.R. Jones
	R.N. Edwards
Chairmen	O.S.A. Sennitt (2015/16)
	D.C. Johnson (2016/17)
Hon. Secretary	D.C.W. Preece
Hon. Treasurer	A.J. Williams

Committee

2013/16	2014/17	2015/18	2016/19
D.J. Phillips	D.C. Johnson	C.F. Dale	M. Tuck
R.T. Griffin	Dr. M.C.A. Puntis	D. Newbery (part)	A.T. Fitzherbert
O.S.A. Sennitt	T.H. Bennett	C.J. Saunders	R. Biles
S.I. Castledine	C.R.V. Norris	S.P. Thorpe	N. Renwick
			D.M. Bevan

Sub-committees

Finance – S.I. Castledine (Ch. 2015/16), C.J. Saunders (Ch. 2016/17), C.V.R. Norris and Hon. Treasurer.

House – D.J. Phillips (Ch. 2015/16), S.P. Thorpe (Ch. 2016/17), Dr. M.C.A. Puntis, R.T. Griffin, M. Tuck, N. Renwick and G. Morgans (co-opted).

Membership – T.H. Bennett (Ch.), C.J. Saunders, S.I. Castledine and A.T. Fitzherbert.

Wine and Social – D.C. Johnson (Ch. 2015/16), C.F. Dale (Ch. 2016/17) C.V.R. Norris, R.D. Gagg, A.P. Embiricos (co-opted), R. Biles, D.M. Bevan and D. Newbery (part).

Art – D.L. Williams (Ch.), D.N. Lermon, D.C. Johnson, A.J. Davies (part), R.J.B. Madley, Dr. P.M. Davies, Dr. P.W. Jones, T.F. Mathias, F. Peel and R.J. Hutchings.

150th – D.C. Johnson (Ch.), Hon. Secretary, J.J. Adams, C.F. Dale, R.D. Gagg, A.P. Embiricos and O.S.A. Sennitt.

6. The Cardiff and County Club's Art Collection

Up until 2001 the Cardiff and County Club had been very fortunate to have works of art loaned by the National Museum of Wales on display on the walls of the Clubhouse in Westgate Street. Their return to the National Museum led to an art committee being established by the then Chairman, David Watson James, plus the creation of a strategy for an art collection, drawn up with the professional help of Peter Wakelin.

It was seen as important that there should be a focus to the collection, rather than just a random selection of artwork. It was agreed to collect works by artists living, or born, in Wales, plus works reflecting the commercial and industrial life of Wales. Since the early 2000s, the Club has therefore obtained numerous works that have made a significant contribution to the character and quality of the Club's premises.

These items have also added to a more diverse body of works acquired previously by the Club and, as of the summer of 2016, there are almost 100 works in the collection as well as a further 20 which have been loaned by members. As a result, the Club has an impressive collection of artwork, and some of the finest pieces have been selected for inclusion in this Members' Supplement to *Always Amongst Friends*.

In addition to the artwork featured in this supplement, there are many other fine works in the Club's collection and, since 2001, some substantial works by Welsh artists of note have been acquired. These include works by Glenys Cour, Arthur Giardelli, Sue Hunt, Will Roberts, Fred Uhlman, Charles White (four) and Ernest Zobole. In addition, there are works of a smaller scale by several artists significant to Wales, including Josef Herman, Shani Rhys James, Donald McIntyre, Leslie Moore, Gwilym Prichard, Will Roberts and Sir Kyffin Williams, as well as prints by Ceri Richards.

'Porth looking towards Pontypridd'

by Denys Short

In the opinion of many members, this is the outstanding work amongst the Club's Collection. This is a major work by an artist which portrays the urban landscape of the industrial valleys, besides representing a central theme in the life of south Wales.

'Duke Street'

by Mark Samuel

This continues the theme of an urban landscape which, given the various locations of the Clubhouse, is so integral to the Club's history and evolution. Duke Street fronts the southern end of the castle and offers an entrance to one of Cardiff's famous Victorian arcades, built in 1902. Mark was born in Kettering and specialises in painting urban landscapes.

Members' Supplement

'City Hall 2011'

by Carl Melegari

Given the close link between the Club and the Bute Estate, it is very fitting that the collection should include artwork of Cardiff's Civic Centre, reputed to be one of the finest in Europe and laid out on land provided by the Marquess following its elevation to city status in the early 1900s. This painting of the City Hall captures the distinctive tone of the Portland Stone and the building's Renaissance style. Born in north Wales and now living in Bristol, Carl paints landscapes and figures in oils in a semi abstract style.

'The Norwegian Church in Cardiff Bay'

by Peter Brown

This was one of the earliest purchases by the Club under the new collection strategy. A painting with a lot of northern light, it celebrates the strong connection between Cardiff and Norway during the days of 'King Coal', when Cardiff was amongst the world's leading coal exporting ports and Norway was a major supplier of timber for pit props and other structures. Although Peter now lives in Bath, he has produced numerous works, painted *en pleine air* in cities such as the Welsh capital.

'Red Scarf'

by Mike Jones

No collection focused on Wales in general and Cardiff in particular would be complete without examples from the mining industry. A number of works in the Club fall into this category, including this piece by Neath-born Mike Jones which shows a miner dressed for his work in the pit wearing a red scarf. Mike now lives in Pontardawe and still takes inspiration from his roots within the mining community.

'Another Great Book'

by Kevin Sinnott

Welsh artist Kevin Sinnott was born in Sarn and has exhibited all around the world, including the Metropolitan Museum of Art in New York, as well as locally at the National Museum of Wales in Cathays Park in Cardiff. His works focus on human relationships and this painting was purchased by the Club in 2001.

Members' Supplement

'Showers in Snowdonia'

by Gareth Parry

Gareth Parry was born in Blaenau Ffestiniog where he worked in the local slate quarry before training at Manchester School of Art and deciding to swap his hammer for a brush. Given his roots in north Wales, he paints atmospheric Welsh landscapes and this painting is a fine example of his art.

'A Tall Waterfall Surrounded by Rugged Hills'

by Peter Prendergast

Peter was born in Abertridwr and initially studied commercial art at Cardiff School of Art, before gaining further experience at the Slade School in London. After his studies, he undertook a number of teaching jobs and moved to north Wales in 1970, where he lived until his death in January 2007.

His works have been exhibited at the National Museum of Wales, The Tate and many other private and public galleries, both at home and abroad.

Members' Supplement

'Open Mountain (after George Dyer)'

by Neil Canning

Neil Canning was born in Enstone, Oxfordshire and spent most of the 1990s living in Wales. This painting was purchased by the Club in 2002, and like most of his works, conjures up memories of distant seas and clear waters.

'Upland Spring'

by David Tress

This is another of the Club's landscape paintings, by London-born David Tress. He took a fine art course at Trent Polytechnic, before moving to Wales in 1976. He has exhibited widely, and was commissioned in 1999 to design one of the Millennium series of stamps for the Royal Mail. His painting style has been referred to as abstract landscape, as shown by this work.